Symmetry in Modeling and Analysis of Dynamic Systems

Symmetry in Modeling and Analysis of Dynamic Systems

Editor

Jan Awrejcewicz

MDPI • Basel • Beijing • Wuhan • Barcelona • Belgrade • Manchester • Tokyo • Cluj • Tianjin

Editor
Jan Awrejcewicz
Department of Automation,
Biomechanics and Mechatronics
Lodz University of Technology
Łódź
Poland

Editorial Office
MDPI
St. Alban-Anlage 66
4052 Basel, Switzerland

This is a reprint of articles from the Special Issue published online in the open access journal *Symmetry* (ISSN 2073-8994) (available at: www.mdpi.com/journal/symmetry/special_issues/Symmetry_Modeling_Analysis_Dynamic_Systems).

For citation purposes, cite each article independently as indicated on the article page online and as indicated below:

LastName, A.A.; LastName, B.B.; LastName, C.C. Article Title. *Journal Name* **Year**, *Volume Number*, Page Range.

ISBN 978-3-0365-3384-1 (Hbk)
ISBN 978-3-0365-3383-4 (PDF)

Contents

About the Editor

Jan Awrejcewicz

Jan Awrejcewicz is Full Professor employed at Lodz University of Technology. He is Head of the Department of Automation, Biomechanics and Mechatronics, and a Full Member of the Polish Academy of Sciences. He has won numerous national/international awards, including The Alexander von Humboldt Award for research and educational achievements as well as honorary doctorates from the Universities in Poland (Bielsko-Biala, Czestochowa, Kielce, Gdańsk), Ukraine (Kharkiv, Dnipro) and Romania (Timisoara).

His research covers mechanics, applied mathematics, material science, biomechanics, automation, physics and computer sciences, with his main focus being nonlinear processes.

To date, he has supervised 27 doctoral students. He is a contributor to 60 different research journals and to 308 conferences, including 80 Keynote/Plenary talks, published 53 monographs (including 27 in English and 1 in Russian) and over 920 papers in scientific journals and conference proceedings.

Preface to "Symmetry in Modeling and Analysis of Dynamic Systems"

Real-world systems exhibit complex behavior, therefore novel mathematical approaches or modifications of classical ones have to be employed to precisely predict, monitor, and control complicated chaotic and stochastic processes. One of the most basic concepts that has to be taken into account while conducting research in all natural sciences is symmetry, and it is usually used to refer to an object that is invariant under some transformations including translation, reflection, rotation or scaling.

The following Special Issue is dedicated to investigations of the concept of dynamical symmetry in the modelling and analysis of dynamic features occurring in various branches of science like physics, chemistry, biology, and engineering, with special emphasis on research based on the mathematical models of nonlinear partial and ordinary differential equations. Addressed topics cover theories developed and employed under the concept of invariance of the global/local behavior of the points of spacetime, including temporal/spatiotemporal symmetries.

Small Solutions of the Perturbed Nonlinear Partial Discrete Dirichlet Boundary Value Problems with (p,q)-Laplacian Operator by Feng Xiong and Zhan Zhou [1] presents results of using critical point theory for study of perturbed partial discrete boundary value problems.

Vibration Properties of a Concrete Structure with Symmetries Used in Civil Engineering by Sorin Vlase, Marin Marin and Ovidiu Deaconu [2] is dedicated to identification of the eigenvalue and eigenmode properties of vibration for components of the concrete constructions' structure allowing for the simplification of their dynamic analysis.

Evaluating the Impact of Different Symmetrical Models of Ambient Assisted Living Systems by Wael Alosaimi, Md Tarique Jamal Ansari, Abdullah Alharbi, Hashem Alyami, Adil Hussain Seh, Abhishek Kumar Pandey, Alka Agrawal and Raees Ahmad Khan [3] provides results of investigation of the potential symmetrical models of Ambient Assisted Living systems and frameworks for the implementation of effective new installations enhancing the living standard for old-aged people.

Dynamical Simulation of Effective Stem Cell Transplantation for Modulation of Microglia Responses in Stroke Treatment by Awatif Jahman Alqarni, Azmin Sham Rambely and Ishak Hashim [4] presents studies of the stability of the mathematical model by using the current biological information on stem cell therapy as a possible treatment for inflammation from microglia during stroke.

In *Second-Order Non-Canonical Neutral Differential Equations with Mixed Type: Oscillatory Behavior* by Osama Moaaz, Amany Nabih, Hammad Alotaibi and Y. S. Hamed [5] new sufficient conditions for the oscillation of solutions of a class of second-order delay differential equations with a mixed neutral term under the non-canonical condition are discussed.

Generalized Attracting Horseshoe in the Rössler Attractor by Karthik Murthy, Ian Jordan, Parth Sojitra, Aminur Rahman and Denis Blackmore [6] presents the generalized attracting horseshoe and its trapping region obtained by using a chosen Poincaré map of the Rössler attractor for an electronic circuit.

Multiple Solutions for a Class of Nonlinear Fourth-Order Boundary Value Problems by Longfei Lin, Yansheng Liu and Daliang Zhao [7] considers multiple solutions for a class of nonlinear fourth-order boundary value problems obtained by constructing a special cone and applying fixed point index theory.

Existence of Three Solutions for a Nonlinear Discrete Boundary Value Problem with ϕ_c-Laplacian by Yanshan Chen and Zhan Zhou [8] is focused on the multiplicity of nontrivial solutions for a nonlinear discrete Dirichlet boundary value problem involving the mean curvature operator.

Last but not least, *On the Absolute Stable Difference Scheme for Third Order Delay Partial Differential Equations* by Allaberen Ashyralyev, Evren Hınçal and Suleiman Ibrahim [9] offers investigations of the third order delay differential equation in a Hilbert space with an unbounded operator.

References

1. Feng Xiong and Zhan Zhou, Small Solutions of the Perturbed Nonlinear Partial Discrete Dirichlet Boundary Value Problems with (p,q)-Laplacian Operator, *Symmetry* 2021, 13(7), 1207;

2. Sorin Vlase, Marin Marin and Ovidiu Deaconu, Vibration Properties of a Concrete Structure with Symmetries Used in Civil Engineering, *Symmetry* 2021, 13(4), 656;

3. Wael Alosaimi, Md Tarique Jamal Ansari, Abdullah Alharbi, Hashem Alyami, Adil Hussain Seh, Abhishek Kumar Pandey, Alka Agrawal and Raees Ahmad Khan

Evaluating the Impact of Different Symmetrical Models of Ambient Assisted Living Systems, *Symmetry* 2021, 13(3), 450;

4. Awatif Jahman Alqarni, Azmin Sham Rambely and Ishak Hashim, Dynamical Simulation of Effective Stem Cell Transplantation for Modulation of Microglia Responses in Stroke Treatment, *Symmetry* 2021, 13(3), 404;

5. Osama Moaaz, Amany Nabih, Hammad Alotaibi and Y. S. Hamed, Second-Order Non-Canonical Neutral Differential Equations with Mixed Type: Oscillatory Behavior, *Symmetry* 2021, 13(2), 318;

6. Karthik Murthy, Ian Jordan, Parth Sojitra, Aminur Rahman and Denis Blackmore, Generalized Attracting Horseshoe in the Rössler Attractor, *Symmetry* 2021, 13(1), 30;

7. Longfei Lin, Yansheng Liu and Daliang Zhao, Multiple Solutions for a Class of Nonlinear Fourth-Order Boundary Value Problems, *Symmetry* 2020, 12(12), 1989;

8. Yanshan Chen and Zhan Zhou, Existence of Three Solutions for a Nonlinear Discrete Boundary Value Problem with ϕ_c-Laplacian, *Symmetry* 2020, 12(11), 1839;

9. Allaberen Ashyralyev, Evren Hınçal and Suleiman Ibrahim, On the Absolute Stable Difference Scheme for Third Order Delay Partial Differential Equations, *Symmetry* 2020, 12(6), 1033.

Jan Awrejcewicz
Editor

Article

Small Solutions of the Perturbed Nonlinear Partial Discrete Dirichlet Boundary Value Problems with (p,q)-Laplacian Operator

Feng Xiong [1,2] and Zhan Zhou [1,2,*]

1 School of Mathematics and Information Science, Guangzhou University, Guangzhou 510006, China; xiongfeng1@e.gzhu.edu.cn
2 Center for Applied Mathematics, Guangzhou University, Guangzhou 510006, China
* Correspondence: zzhou0321@hotmail.com

Abstract: In this paper, we consider a perturbed partial discrete Dirichlet problem with the (p,q)-Laplacian operator. Using critical point theory, we study the existence of infinitely many small solutions of boundary value problems. Without imposing the symmetry at the origin on the nonlinear term f, we obtain the sufficient conditions for the existence of infinitely many small solutions. As far as we know, this is the study of perturbed partial discrete boundary value problems. Finally, the results are exemplified by an example.

Keywords: boundary value problem; partial difference equation; infinitely many small solutions; (p,q)-Laplacian; critical point theory

Citation: Xiong, F.; Zhou, Z. Small Solutions of the Perturbed Nonlinear Partial Discrete Dirichlet Boundary Value Problems with (p,q)-Laplacian Operator. *Symmetry* **2021**, *13*, 1207. https://doi.org/10.3390/sym13071207

Academic Editor: Jan Awrejcewicz

Received: 25 May 2021
Accepted: 3 July 2021
Published: 5 July 2021

1. Introduction

Let $\mathbf{Z}$ and $\mathbf{R}$ denote the sets of integers and real numbers, respectively. Denote $\mathbf{Z}(a,b)=\{a,a+1,\cdots,b\}$ when $a\le b$.

We consider the following partial discrete problem, namely $(D^{\lambda,\mu})$
$-[\Delta_1(\phi_p(\Delta_1 y(s-1,t)))+\Delta_2(\phi_p(\Delta_2 y(s,t-1)))]+l(s,t)\phi_q(y(s,t))=\lambda f((s,t),y(s,t))$ $+\mu g((s,t),y(s,t)),(s,t)\in\mathbf{Z}(1,a)\times\mathbf{Z}(1,b)$, with Dirichlet boundary conditions as follows:

$$\begin{aligned} y(s,0)&=y(s,b+1)=0, \quad s\in\mathbf{Z}(0,a+1),\\ y(0,t)&=y(a+1,t)=0, \quad t\in\mathbf{Z}(0,b+1), \end{aligned} \tag{1}$$

where a and b are the given positive integers, λ and μ are the positive real parameters, Δ_1 and Δ_2 are the forward difference operators defined by $\Delta_1 y(s,t)=y(s+1,t)-y(s,t)$ and $\Delta_2 y(s,t)=y(s,t+1)-y(s,t)$, $\Delta_1^2 y(s,t)=\Delta_1(\Delta_1 y(s,t))$ and $\Delta_2^2 y(s,t)=\Delta_2(\Delta_2 y(s,t))$, $\phi_r(y)=|y|^{r-2}y$ with $y\in\mathbf{R}$, $1<q\le p<+\infty$, $l(s,t)\ge 0$ for all $(s,t)\in\mathbf{Z}(1,a)\times\mathbf{Z}(1,b)$, and $f((s,t),\cdot),g((s,t),\cdot)\in C(\mathbf{R},\mathbf{R})$ for each $(s,t)\in\mathbf{Z}(1,a)\times\mathbf{Z}(1,b)$.

Difference equations are widely applied in diverse domains, including natural science, and biological neural networks, as shown in [1–4]. For the existence and multiplicity of solutions to boundary value problems, some authors derived a number of conclusions using nonlinear analysis methods, such as fixed point methods as well as the Brouwer degree [5–9]. In 2003, Yu and Guo [10] used firstly the critical point theory to study a class of difference equations. Since then, many mathematical researchers have explored difference equations and made great achievements, which include the results of periodic solutions [10,11], homoclinic solutions [12–18], boundary value problems [19–25] and so on.

Bonanno et al. [20] in 2016 considered the following discrete Dirichlet problem:

$$\begin{cases} \Delta^2 y_{h-1}+\lambda f(h,y_h)=0, & h\in\mathbf{Z}(1,N),\\ y_0=y_{N+1}=0, \end{cases} \tag{2}$$

and acquired at least two positive solutions of (2).

Mawhin et al. [21] in 2017 studied the following boundary value problem:

$$\begin{cases} -\Delta(\phi_p(\Delta y_{h-1})) + q_h\phi_p(y_h) = \lambda f(h, y_h), & h \in \mathbf{Z}(1, N), \\ y_0 = y_{N+1} = 0, \end{cases} \tag{3}$$

extending the results in [20] with $p = 2$.

Nastasi et al. [22] in 2017 studied the discrete Dirichlet problem involving the (p, q)-Laplacian operator as follows:

$$\begin{cases} -\Delta(\phi_p(\Delta y(h-1))) - \Delta(\phi_q(\Delta y(h-1))) + \alpha(h)\phi_p(y(h)) + \beta(h)\phi_q(y(h)) = \lambda g(h, y(h)), \ h \in \mathbf{Z}(1, N), \\ y(0) = y(N+1) = 0, \end{cases} \tag{4}$$

and obtained at least two positive solutions of (4).

In 2019, Ling and Zhou [24] considered the Dirichlet problem involving ϕ_c-Laplacian as follows:

$$\begin{cases} -\Delta(\phi_c(\Delta y_{h-1})) + q_h\phi_c(y_h) = \lambda f(h, y_h), & h \in \mathbf{Z}(1, N), \\ y_0 = y_{N+1} = 0. \end{cases} \tag{5}$$

They studied the existence of positive solutions of (5) when $q_h \equiv 0$ in [23].

In 2020, Wang and Zhou [25] considered discrete Dirichlet boundary value problem as follows:

$$\begin{cases} \Delta(\phi_{p,c}(\Delta y_{h-1})) + \lambda f(h, y_h) = 0, & h \in \mathbf{Z}(1, N), \\ y_0 = y_{N+1} = 0. \end{cases} \tag{6}$$

The difference equations studied above involve only one variable. However, the difference equations containing two or more variables are less studied, and such difference equations are called partial difference equations. Recently, partial difference equations were widely used in many fields. Boundary value problems of partial difference equations seem to be challenging problem that has attracted many mathematical researchers [26,27].

In 2015, Heidarkhani and Imbesi [26] adopted two critical points theorems to establish multiple solutions of the partial discrete problem as shown below:

$$\Delta_1^2 y(s-1, t) + \Delta_2^2 y(s, t-1) + \lambda f((s, t), y(s, t)) = 0, \quad (s, t) \in \mathbf{Z}(1, a) \times \mathbf{Z}(1, b), \tag{7}$$

with Dirichlet boundary conditions (1).

Recently, in 2020, Du and Zhou [27] studied a partial discrete Dirichlet problem as follows:

$$\Delta_1(\phi_p(\Delta_1 y(s-1, t))) + \Delta_2(\phi_p(\Delta_2 y(s, t-1))) + \lambda f((s, t), y(s, t)) = 0, \quad (s, t) \in \mathbf{Z}(1, a) \times \mathbf{Z}(1, b), \tag{8}$$

with Dirichlet boundary conditions (1).

Inspired by the above research, we found that the perturbed partial difference equations had rarely been studied, so this paper aims at studying small solutions of the perturbed partial discrete Dirichlet problems with the (p, q)-Laplacian operator. Here, the perturbed partial difference equations mean that the term with the parameter μ in the right hand of the equation for the problem $(D^{\lambda,\mu})$ is very small. A solution $y(s, t)$ of $(D^{\lambda,\mu})$ is called a small solution if the norm $\|y(s, t)\|$ is small. In fact, without the symmetric assumption on the origin for the nonlinear term f, we can still verify that problem $(D^{\lambda,\mu})$ possesses a sequence of solutions which converges to zero by using the Lemma 2. Moreover, by Lemma 1, we can show that all of these solutions are positive. Furthermore, by truncation techniques, we obtain two sequences of constant-sign solutions, which converge to zero (with one being positive and the other being negative). As far as we know, our study takes the lead in addressing small solutions of the perturbed partial discrete Dirichlet problems with the (p, q)-Laplacian operator.

The rest of this paper is organized as follows. In Section 2, we establish the variational framework linked to $(D^{\lambda,\mu})$ and recall the abstract critical point theorem. In Section 3, we give the main results. In Section 4, we provide an example to demonstrate our results. We make a conclusion in the last section.

2. Preliminaries

The current section is the first one to establish the variational framework linked to $(D^{\lambda,\mu})$. We consider the *ab*-dimensional Banach space

$Y = \{y : \mathbf{Z}(0, a+1) \times \mathbf{Z}(0, b+1) \to \mathbf{R} : y(s, 0) = y(s, b+1) = 0, s \in \mathbf{Z}(0, a+1)$ and $y(0, t) = y(a+1, t) = 0, t \in \mathbf{Z}(0, b+1)\}$, endowed with the norm

$$\|y\| = \left(\sum_{t=1}^{b} \sum_{s=1}^{a+1} |\Delta_1 y(s-1, t)|^p + \sum_{s=1}^{a} \sum_{t=1}^{b+1} |\Delta_2 y(s, t-1)|^p \right)^{\frac{1}{p}}, \quad y \in Y.$$

and $\|y\|_\infty = \max\{|y(s,t)| : (s,t) \in \mathbf{Z}(1,a) \times \mathbf{Z}(1,b)\}$ is another norm in Y.

Let $l_* = \min\{l(s,t) : (s,t) \in \mathbf{Z}(1,a) \times \mathbf{Z}(1,b)\}$.

Proposition 1. *The following inequality holds:*

$$\|y\|_\infty \leq \max\left\{ \left(\frac{p(a+b+2)^{p-1}}{4^p + l_*(a+b+2)^{p-1}} \right)^{1/q} \left(\frac{\|y\|^p}{p} + \frac{\sum_{t=1}^{b} \sum_{s=1}^{a} l(s,t)|y(s,t)|^q}{q} \right)^{1/q}, \right.$$
$$\left. \left(\frac{p(a+b+2)^{p-1}}{4^p + l_*(a+b+2)^{p-1}} \right)^{1/p} \left(\frac{\|y\|^p}{p} + \frac{\sum_{t=1}^{b} \sum_{s=1}^{a} l(s,t)|y(s,t)|^q}{q} \right)^{1/p} \right\}.$$

Proof. According to the result of ([27], Proposition 1), we have the following:

$$\|y\|_\infty^p \leq \frac{(a+b+2)^{p-1}}{4^p} \|y\|^p. \tag{9}$$

When $\|y\|_\infty > 1$, according to (9), we have the following:

$$\begin{aligned}
\frac{(1 + l_* \frac{(a+b+2)^{p-1}}{4^p}) \|y\|_\infty^q}{p} &\leq \frac{\|y\|_\infty^p + \frac{(a+b+2)^{p-1}}{4^p} \sum_{t=1}^{b} \sum_{s=1}^{a} l(s,t)|y(s,t)|^q}{p} \\
&\leq \frac{(a+b+2)^{p-1}}{p4^p} \left(\|y\|^p + \sum_{t=1}^{b} \sum_{s=1}^{a} l(s,t)|y(s,t)|^q \right) \\
&\leq \frac{(a+b+2)^{p-1}}{p4^p} \|y\|^p + \frac{(a+b+2)^{p-1}}{q4^p} \sum_{t=1}^{b} \sum_{s=1}^{a} l(s,t)|y(s,t)|^q,
\end{aligned} \tag{10}$$

that is,

$$\|y\|_\infty \leq \left(\frac{p(a+b+2)^{p-1}}{4^p + l_*(a+b+2)^{p-1}} \right)^{1/q} \left(\frac{\|y\|^p}{p} + \frac{\sum_{t=1}^{b} \sum_{s=1}^{a} l(s,t)|y(s,t)|^q}{q} \right)^{1/q}. \tag{11}$$

When $\|y\|_\infty \leq 1$, according to (9), we have the following:

$$\frac{(1+l_*\frac{(a+b+2)^{p-1}}{4^p})\|y\|_\infty^p}{p} \leq \frac{\|y\|_\infty^p + \frac{(a+b+2)^{p-1}}{4^p}\sum\limits_{t=1}^{b}\sum\limits_{s=1}^{a} l(s,t)|y(s,t)|^q}{p} \leq \frac{(a+b+2)^{p-1}}{p4^p}\|y\|^p + \frac{(a+b+2)^{p-1}}{q4^p}\sum_{t=1}^{b}\sum_{s=1}^{a} l(s,t)|y(s,t)|^q, \tag{12}$$

that is,

$$\|y\|_\infty \leq \left(\frac{p(a+b+2)^{p-1}}{4^p + l_*(a+b+2)^{p-1}}\right)^{1/p}\left(\frac{\|y\|^p}{p} + \frac{\sum\limits_{t=1}^{b}\sum\limits_{s=1}^{a} l(s,t)|y(s,t)|^q}{q}\right)^{1/p}. \tag{13}$$

In summary, we have the following:

$$\|y\|_\infty \leq \max\left\{\left(\frac{p(a+b+2)^{p-1}}{4^p + l_*(a+b+2)^{p-1}}\right)^{1/q}\left(\frac{\|y\|^p}{p} + \frac{\sum\limits_{t=1}^{b}\sum\limits_{s=1}^{a} l(s,t)|y(s,t)|^q}{q}\right)^{1/q}, \left(\frac{p(a+b+2)^{p-1}}{4^p + l_*(a+b+2)^{p-1}}\right)^{1/p}\left(\frac{\|y\|^p}{p} + \frac{\sum\limits_{t=1}^{b}\sum\limits_{s=1}^{a} l(s,t)|y(s,t)|^q}{q}\right)^{1/p}\right\}.$$

Define

$$\Phi(y) = \Phi_1(y) + \Phi_2(y),$$

$$\Psi(y) = \sum_{t=1}^{b}\sum_{s=1}^{a}\Big(F((s,t),y(s,t)) + \frac{\mu}{\lambda}G((s,t),y(s,t))\Big),$$

for every $y \in Y$, where $\Phi_1(y) = \frac{\|y\|^p}{p}$, $\Phi_2(y) = \frac{\sum\limits_{t=1}^{b}\sum\limits_{s=1}^{a} l(s,t)|y(s,t)|^q}{q}$, $F((s,t),y) = \int_0^y f((s,t),\tau)\, d\tau$, $G((s,t),y) = \int_0^y g((s,t),\tau)d\tau$ for each $((s,t),y) \in \mathbf{Z}(1,a) \times \mathbf{Z}(1,b) \times \mathbf{R}$.

Let

$$I_\lambda(y) = \Phi(y) - \lambda\Psi(y),$$

for any $y \in Y$. Obviously, $\Phi, \Psi \in C^1(Y,R)$, that is, Φ_1, Φ_2 and Ψ are continuously Fréchet differentiable in Y.

$$\begin{aligned}
\Phi_1'(y)(v) &= \lim_{t\to 0}\frac{\Phi_1(y+tv) - \Phi_1(y)}{t} \\
&= \sum_{t=1}^{b}\sum_{s=1}^{a+1}\phi_p(\Delta_1 y(s-1,t))\Delta_1 v(s-1,t) + \sum_{s=1}^{a}\sum_{t=1}^{b+1}\phi_p(\Delta_2 y(s,t-1))\Delta_2 v(s,t-1) \\
&= -\sum_{t=1}^{b}\sum_{s=1}^{a}\Delta_1\phi_p(\Delta_1 y(s-1,t))v(s,t) - \sum_{s=1}^{a}\sum_{t=1}^{b}\Delta_2\phi_p(\Delta_2 y(s,t-1))v(s,t),
\end{aligned}$$

$$\Phi_2'(y)(v) = \sum_{t=1}^{b}\sum_{s=1}^{a} l(s,t)\phi_q(y(s,t))v(s,t),$$

and

$$\Psi'(y)(v) = \lim_{t\to 0}\frac{\Psi(y+tv)-\Psi(y)}{t} = \sum_{t=1}^{b}\sum_{s=1}^{a}\Big(f((s,t),y(s,t)) + \frac{\mu}{\lambda}g((s,t),y(s,t))\Big)v(s,t),$$

for $y, v \in Y$.

Thus

$$\begin{aligned}[\Phi'(y) - \lambda\Psi'(y)](v) = -\sum_{t=1}^{b}\sum_{s=1}^{a}\{&\Delta_1\phi_p(\Delta_1 y(s-1,t)) + \Delta_2\phi_p(\Delta_2 y(s,t-1)) \\ &- l(s,t)\phi_q(y(s,t)) + \lambda f((s,t),y(s,t)) + \mu g((s,t),y(s,t))\}v(s,t) = 0, \quad \forall v(s,t) \in Y,\end{aligned} \tag{14}$$

is equivalent to

$$-\Big[\Delta_1(\phi_p(\Delta_1 y(s-1,t))) + \Delta_2(\phi_p(\Delta_2 y(s,t-1)))\Big] + l(s,t)\phi_q(y(s,t)) = \lambda f((s,t),y(s,t)) +$$

$\mu g((s,t),y(s,t))$, for any $(s,t) \in \mathbf{Z}(1,a)\times\mathbf{Z}(1,b)$ with $y(s,0) = y(s,b+1) = 0$, $s \in \mathbf{Z}(0,a+1)$, $y(0,t) = y(a+1,t) = 0, t \in \mathbf{Z}(0,b+1)$. Thus, we reduce the existence of the solutions of $(D^{\lambda,\mu})$ to the existence of the critical points of $\Phi - \lambda\Psi$ on Y. □

Lemma 1. *Suppose that there exists y: $\mathbf{Z}(0,a+1)\times\mathbf{Z}(0,b+1)\to\mathbf{R}$ such that the following is true:*

$$y(s,t) > 0 \ \ or \ \ -\Delta_1(\phi_p(\Delta_1 y(s-1,t))) - \Delta_2(\phi_p(\Delta_2 y(s,t-1))) + l(s,t)\phi_q(y(s,t)) \geq 0, \tag{15}$$

for all $(s,t) \in \mathbf{Z}(1,a)\times\mathbf{Z}(1,b)$ and $y(s,0) = y(s,b+1) = 0, s \in \mathbf{Z}(0,a+1)$, $y(0,t) = y(a+1,t) = 0$, $t \in \mathbf{Z}(0,b+1)$.

Then, either $y(s,t) > 0$ for all $(s,t) \in \mathbf{Z}(1,a)\times\mathbf{Z}(1,b)$ or $y \equiv 0$.

Proof. Let $h \in \mathbf{Z}(1,a), k \in \mathbf{Z}(1,b)$ and

$$y(h,k) = \min\{y(s,t) : s \in \mathbf{Z}(1,a), t \in \mathbf{Z}(1,b)\}.$$

If $y(h,k) > 0$, then it is clear that $y(s,t) > 0$ for all $s \in \mathbf{Z}(1,a), t \in \mathbf{Z}(1,b)$. If $y(h,k) \leq 0$, then $y(h,k) = \min\{y(s,t) : s \in \mathbf{Z}(0,a+1), t \in \mathbf{Z}(0,b+1)\}$, since $\Delta_1 y(h-1,k) = y(h,k) - y(h-1,k) \leq 0$, $\Delta_2 y(h,k-1) = y(h,k) - y(h,k-1) \leq 0$, and $\Delta_1 y(h,k) = y(h+1,k) - y(h,k) \geq 0$, $\Delta_2 y(h,k) = y(h,k+1) - y(h,k) \geq 0$, $\phi_p(s)$ is increasing in s, and $\phi_p(0) = 0$, we have

$$\phi_p(\Delta_1 y(h,k)) \geq 0 \geq \phi_p(\Delta_1 y(h-1,k)),$$

and

$$\phi_p(\Delta_2 y(h,k)) \geq 0 \geq \phi_p(\Delta_2 y(h,k-1)).$$

Owing to $\Delta_1(\phi_p(\Delta_1 y(h-1,k))) = \phi_p(\Delta_1 y(h,k)) - \phi_p(\Delta_1 y(h-1,k)) \geq 0$, $\Delta_2(\phi_p(\Delta_2 y(h,k-1))) = \phi_p(\Delta_2 y(h,k)) - \phi_p(\Delta_2 y(h,k-1)) \geq 0$. Thus, we have

$$\Delta_1(\phi_p(\Delta_1 y(h-1,k))) + \Delta_2(\phi_p(\Delta_2 y(h,k-1))) \geq 0. \tag{16}$$

By (15), we have the following:

$$-\Delta_1(\phi_p(\Delta_1 y(h-1,k))) - \Delta_2(\phi_p(\Delta_2 y(h,k-1))) \geq -l(h,k)\phi_q(y(h,k)) \geq 0,$$

that is

$$\Delta_1(\phi_p(\Delta_1 y(h-1,k))) + \Delta_2(\phi_p(\Delta_2 y(h,k-1))) \leq 0. \tag{17}$$

By combining (16) with (17), we have the following:

$$\Delta_1(\phi_p(\Delta_1 y(h-1,k))) + \Delta_2(\phi_p(\Delta_2 y(h,k-1))) = 0,$$

namely,

$$\begin{cases} \phi_p(\Delta_1 y(h,k)) = \phi_p(\Delta_1 y(h-1,k)) = 0, \\ \phi_p(\Delta_2 y(h,k)) = \phi_p(\Delta_2 y(h,k-1)) = 0. \end{cases}$$

Therefore,

$$\begin{cases} y(h+1,k) = y(h,k) = y(h-1,k), \\ y(h,k+1) = y(h,k) = y(h,k-1). \end{cases}$$

If $h+1 = a+1$, we have $y(h,k) = 0$. Otherwise, $(h+1) \in \mathbf{Z}(1,a)$. Replacing h by $h+1$, we get $y(h+2,k) = y(h+1,k)$. Continuing this process $(a+1-h)$ times, we have $y(h,k) = y(h+1,k) = y(h+2,k) = \cdots = y(a+1,k) = 0$. Similarly, we have $y(h,k) = y(h-1,k) = y(h-2,k) = \cdots = y(0,k) = 0$. Hence, $y(s,k) = 0$ for each $s \in \mathbf{Z}(1,a)$. In the same way, we can prove that $y \equiv 0$, and the proof is completed. □

From Lemma 1, we have the following:

Corollary 1. *Suppose that there exists y: $\mathbf{Z}(0,a+1) \times \mathbf{Z}(0,b+1) \to \mathbf{R}$ such that*

$$y(s,t) < 0 \ \ or \ \ -\Delta_1(\phi_p(\Delta_1 y(s-1,t))) - \Delta_2(\phi_p(\Delta_2 y(s,t-1))) + l(s,t)\phi_q(y(s,t)) \le 0, \quad (18)$$

for all $(s,t) \in \mathbf{Z}(1,a) \times \mathbf{Z}(1,b)$ and $y(s,0) = y(s,b+1) = 0, s \in \mathbf{Z}(0,a+1)$, $y(0,t) = y(a+1,t) = 0, t \in \mathbf{Z}(0,b+1)$.

Then, either $y(s,t) < 0$ for all $(s,t) \in \mathbf{Z}(1,a) \times \mathbf{Z}(1,b)$ or $y \equiv 0$.

The existence of constant-sign solutions is discussed by truncation techniques. So, we introduce the following truncations of the functions $f((s,t),\xi)$ and $g((s,t),\xi)$ for every $(s,t) \in \mathbf{Z}(1,a) \times \mathbf{Z}(1,b)$.

If $f((s,t),0) \ge 0$ and $g((s,t),0) \ge 0$ for every $(s,t) \in \mathbf{Z}(1,a) \times \mathbf{Z}(1,b)$, let

$$f^+((s,t),\xi) := \begin{cases} f((s,t),\xi), & \text{if } \xi \ge 0, \\ f((s,t),0), & \text{if } \xi < 0, \end{cases} \quad g^+((s,t),\xi) := \begin{cases} g((s,t),\xi), & \text{if } \xi \ge 0, \\ g((s,t),0), & \text{if } \xi < 0. \end{cases}$$

Define problem (D^{λ,μ^+}) as follows:

$-\Big[\Delta_1(\phi_p(\Delta_1 y(s-1,t))) + \Delta_2(\phi_p(\Delta_2 y(s,t-1)))\Big] + l(s,t)\phi_q(y(s,t)) = \lambda f^+((s,t), y(s,t)) + \mu g^+((s,t),y(s,t)), (s,t) \in \mathbf{Z}(1,a) \times \mathbf{Z}(1,b)$, with Dirichlet boundary conditions (1).

Obviously, $f^+((s,t),\cdot)$ and $g^+((s,t),\cdot)$ are also continuous for every $(s,t) \in \mathbf{Z}(1,a) \times \mathbf{Z}(1,b)$. By Lemma 1, the solutions of problem (D^{λ,μ^+}) are also those of problem $(D^{\lambda,\mu})$. Therefore, when problem (D^{λ,μ^+}) has non-zero solutions, then problem $(D^{\lambda,\mu})$ possesses positive solutions.

If $f((s,t),0) \le 0$ and $g((s,t),0) \le 0$ for every $(s,t) \in \mathbf{Z}(1,a) \times \mathbf{Z}(1,b)$, let

$$f^-((s,t),\xi) := \begin{cases} f((s,t),0), & \text{if } \xi > 0, \\ f((s,t),\xi), & \text{if } \xi \le 0, \end{cases} \quad g^-((s,t),\xi) := \begin{cases} g((s,t),0), & \text{if } \xi > 0, \\ g((s,t),\xi), & \text{if } \xi \le 0. \end{cases}$$

Define problem (D^{λ,μ^-}) as follows:

$-\Big[\Delta_1(\phi_p(\Delta_1 y(s-1,t))) + \Delta_2(\phi_p(\Delta_2 y(s,t-1)))\Big] + l(s,t)\phi_q(y(s,t)) = \lambda f^-((s,t), y(s,t)) + \mu g^-((s,t),y(s,t)), (s,t) \in \mathbf{Z}(1,a) \times \mathbf{Z}(1,b)$, with Dirichlet boundary conditions (1).

By Corollary 1, the solutions of problem (D^{λ,μ^-}) are also those of problem $(D^{\lambda,\mu})$. Therefore, when problem (D^{λ,μ^-}) has non-zero solutions, then problem $(D^{\lambda,\mu})$ possesses negative solutions.

Here, we present the main tools used in this paper.

Lemma 2 (Theorem 4.3 of [28]). *Let X be a finite dimensional Banach space and let $I_\lambda : X \to \mathbf{R}$ be a function satisfying the following structure hypothesis:*

(H) $I_\lambda(u) := \Phi(u) - \lambda\Psi(u)$ for all $u \in X$, where $\Phi, \Psi : X \to \mathbf{R}$ be two continuously Gâteux differentiable functions with Φ coercive, i.e., $\lim_{\|u\|\to+\infty} \Phi(u) = +\infty$, and such that $\inf_X \Phi = \Phi(0) = \Psi(0) = 0$.

For all $r > 0$, put the following:

$$\varphi(r) := \frac{\sup_{v\in\Phi^{-1}[0,r]} \Psi(v)}{r}, \quad \varphi_0 := \liminf_{r\to 0^+} \varphi(r).$$

Assume that $\varphi_0 < +\infty$ and for every $\lambda \in (0, \frac{1}{\varphi_0})$, 0 is not a local minima of functional I_λ. Then, there is a sequence $\{u_n\}$ of pairwise distinct critical points (local minima) of I_λ such that $\lim_{n\to+\infty} u_n = 0$.

3. Main Results

In this section, the existence of constant-sign solutions of problem $(D^{\lambda,\mu})$ is discussed. Our aim is to use Lemma 2 for the function $I_\lambda^\pm : X \to R$,

$$I_\lambda^\pm(y) := \Phi(y) - \lambda\Psi^\pm(y),$$

where

$$\Psi^\pm(y) = \sum_{t=1}^{b}\sum_{s=1}^{a}(F^\pm((s,t),y(s,t)) + \frac{\mu}{\lambda}G^\pm((s,t),y(s,t))),$$

$$F^\pm((s,t),y) := \int_0^y f^\pm((s,t),\tau)d\tau, \quad G^\pm((s,t),y) := \int_0^y g^\pm((s,t),\tau)d\tau,$$

for each $(s,t) \in \mathbf{Z}(1,a) \times \mathbf{Z}(1,b)$. Then, we apply Lemma 1 or Corollary 1 to obtain our results.

Let

$$A_{0*} = \liminf_{c\to 0^+} \frac{\sum_{t=1}^{b}\sum_{s=1}^{a} \max_{0\le m\le c} F((s,t),\pm m)}{c^p}, \quad B^{0*} = \limsup_{c\to 0^+} \frac{\sum_{t=1}^{b}\sum_{s=1}^{a} F((s,t),c)}{c^p}.$$

$$C^{0*} = \limsup_{c\to 0^+} \frac{\sum_{t=1}^{b}\sum_{s=1}^{a} \max_{0\le m\le c} G((s,t),\pm m)}{c^p}, \quad \tilde{l} = \sum_{t=1}^{b}\sum_{s=1}^{a} l(s,t).$$

In addition, put the following:

$$\bar{\mu}_\lambda^* := \frac{1}{C^{0*}}\left(\frac{4^p + l_*(a+b+2)^{p-1}}{p(a+b+2)^{p-1}} - \lambda A_{0*}\right).$$

It should be pointed out that if the denominator is 0, we regard $\frac{1}{0}$ as $+\infty$.

Theorem 1. *Let $f((s,t),y)$ be a continuous function of y, and $f((s,t),0) \ge 0$, $g((s,t),\cdot) \in C(\mathbf{R},\mathbf{R})$ for every $(s,t) \in \mathbf{Z}(1,a) \times \mathbf{Z}(1,b)$. Suppose the following:*

(i_1) $\frac{p(a+b+2)^{p-1}A_{0+}}{4^p + l_*(a+b+2)^{p-1}} < \frac{pqB^{0^+}}{2bq+2aq+p\tilde{l}}$,

(g_1) there exsits $\delta > 0$ such that at $[0,\delta]$, $G((s,t),y) \ge 0$ and $C^{0^+} < +\infty$.

Then, for each $\lambda \in \Lambda = \left(\frac{2bq+2aq+p\tilde{l}}{pqB^{0^+}}, \frac{4^p+l_*(a+b+2)^{p-1}}{p(a+b+2)^{p-1}A_{0+}}\right)$ and $\mu \in [0, \bar{\mu}_\lambda^+)$, problem $(D^{\lambda,\mu})$ has a sequence of positive solutions, which converges to zero.

Proof. We take $X = Y$, Φ_1, Φ_2 and Ψ as in Section 2. Obviously, for each $(s,t) \in \mathbf{Z}(1,a) \times \mathbf{Z}(1,b)$, $g((s,t),0) \ge 0$.

Now, we consider the auxiliary problem (D^{λ,μ^+}).
Clearly Φ and Ψ^+ satisfy the hypothesis required in Lemma 2.
Let

$$r = \frac{4^p + l_*(a+b+2)^{p-1}}{p(a+b+2)^{p-1}} \min\{c^q, c^p\}, \quad \text{for } c > 0.$$

Assume $y \in Y$, and the following:

$$\Phi(y) = \frac{\|y\|^p}{p} + \frac{\sum\limits_{t=1}^{b}\sum\limits_{s=1}^{a} l(s,t)|y(s,t)|^q}{q} \leq r.$$

If $r = \frac{4^p+l_*(a+b+2)^{p-1}}{p(a+b+2)^{p-1}}c^q$, it means that $c \geq 1$. According to Proposition 1, we have the following:

$$\begin{aligned}\|y\|_\infty &\leq \max\left\{\left(\frac{p(a+b+2)^{p-1}}{4^p + l_*(a+b+2)^{p-1}}\right)^{1/q} r^{1/q},\right.\\ &\left.\left(\frac{p(a+b+2)^{p-1}}{4^p + l_*(a+b+2)^{p-1}}\right)^{1/p} r^{1/p}\right\}\\ &= \max\{c, c^{q/p}\} = c.\end{aligned}$$

If $r = \frac{4^p+l_*(a+b+2)^{p-1}}{p(a+b+2)^{p-1}}c^p$, we know $0 < c < 1$, then

$$\begin{aligned}\|y\|_\infty &\leq \max\left\{\left(\frac{p(a+b+2)^{p-1}}{4^p + l_*(a+b+2)^{p-1}}\right)^{1/q} r^{1/q},\right.\\ &\left.\left(\frac{p(a+b+2)^{p-1}}{4^p + l_*(a+b+2)^{p-1}}\right)^{1/p} r^{1/p}\right\}\\ &= \max\{c^{p/q}, c\} = c,\end{aligned}$$

and

$$\frac{c^p}{r} = \frac{p(a+b+2)^{p-1}}{4^p + l_*(a+b+2)^{p-1}}. \tag{19}$$

Therefore, we have $\Phi^{-1}[0,r] \subseteq \{y \in Y : \|y\|_\infty \leq c\}$.
By the definition of φ, we have the following:

$$\begin{aligned}\varphi(r) &= \frac{\sup\limits_{v\in\Phi^{-1}[0,r]} \Psi^+(v)}{r}\\ &\leq \frac{1}{r}\sup_{\|y\|_\infty \leq c}\sum_{t=1}^{b}\sum_{s=1}^{a}\left(F^+((s,t),y(s,t)) + \frac{\mu}{\lambda}G^+((s,t),y(s,t))\right)\\ &\leq \frac{c^p}{r}\left(\frac{\sum\limits_{t=1}^{b}\sum\limits_{s=1}^{a}\max\limits_{0\leq m\leq c} F((s,t),m)}{c^p} + \frac{\mu}{\lambda}\frac{\sum\limits_{t=1}^{b}\sum\limits_{s=1}^{a}\max\limits_{0\leq m\leq c} G((s,t),m)}{c^p}\right).\end{aligned}$$

According to condition (i_1), (g_1) and (19), we have the following:

$$\varphi_0 \leq \frac{p(a+b+2)^{p-1}}{4^p + l_*(a+b+2)^{p-1}}(A_{0^+} + \frac{\mu}{\lambda}C^{0^+}) < +\infty.$$

We assert that if $\lambda \in \left(\frac{2bq+2aq+p\tilde{l}}{pqB^{0^+}}, \frac{4^p+l_*(a+b+2)^{p-1}}{p(a+b+2)^{p-1}A_{0^+}}\right)$ and $\mu \in [0, \bar{\mu}^+_\lambda)$, then $\lambda \in (0, \frac{1}{\varphi_0})$.

In fact, for $\lambda \in \Lambda$, we have $\lambda > 0$.

When $C^{0^+} = 0$, then

$$\varphi_0 \leq \frac{p(a+b+2)^{p-1}}{4^p + l_*(a+b+2)^{p-1}} A_{0^+} < \frac{1}{\lambda}.$$

When $C^{0^+} > 0$, then

$$\begin{aligned}\varphi_0 &< \frac{p(a+b+2)^{p-1}}{4^p + l_*(a+b+2)^{p-1}} \left(A_{0^+} + \frac{\bar{\mu}_\lambda^+}{\lambda} C^{0^+} \right) \\ &= \frac{p(a+b+2)^{p-1}}{4^p + l_*(a+b+2)^{p-1}} \left(A_{0^+} + \frac{1}{\lambda} \frac{1}{C^{0^+}} \left(\frac{4^p + l_*(a+b+2)^{p-1}}{p(a+b+2)^{p-1}} - \lambda A_{0^+} \right) C^{0^+} \right) \\ &= \frac{1}{\lambda}.\end{aligned}$$

Clearly, $(0, 0, \cdots, 0) \in Y$ is a global minima of Φ.

Next, we need to prove that $(0, 0, \cdots, 0)$ is not a local minima of I_λ^+. Let us prove this in two cases: $B^{0^+} = +\infty$ and $B^{0^+} < +\infty$.

Firstly, when $B^{0^+} = +\infty$, fix M such that $M > \frac{2aq+2bq+\tilde{l}p}{pq}$ and there exists a sequence of positive numbers $\{c_n\}$ such that $\lim_{n\to+\infty} c_n = 0$, and

$$\sum_{t=1}^{b} \sum_{s=1}^{a} F^+((s,t), c_n) = \sum_{t=1}^{b} \sum_{s=1}^{a} F((s,t), c_n) \geq \frac{M c_n^q}{\lambda}, \quad \text{for } n \in \mathbf{Z}(1).$$

Define a sequence $\{\eta_n\}$ in Y with the following:

$$\eta_n(s,t) = \begin{cases} c_n, & \text{if } (s,t) \in \mathbf{Z}(1,a) \times \mathbf{Z}(1,b), \\ 0, & \text{if } s = 0, t \in \mathbf{Z}(0, b+1), \text{ or, } s = a+1, t \in \mathbf{Z}(0, b+1), \\ 0, & \text{if } t = 0, s \in \mathbf{Z}(0, a+1), \text{ or, } t = b+1, s \in \mathbf{Z}(0, a+1). \end{cases}$$

According to $G^+((s,t), \eta_n(s,t)) = G((s,t), \eta_n(s,t)) \geq 0$, $(s,t) \in \mathbf{Z}(1,a) \times \mathbf{Z}(1,b)$, we acquire the following:

$$\begin{aligned} I_\lambda^+(\eta_n) &\leq \left(\frac{2a+2b}{p} \right) c_n^p + \frac{\tilde{l}}{q} c_n^q - \lambda \left(\sum_{t=1}^{b} \sum_{s=1}^{a} F((s,t), c_n) \right) \\ &\leq \left(\frac{2a+2b}{p} \right) c_n^q + \frac{\tilde{l}}{q} c_n^q - M c_n^q \\ &= \left(\frac{2a+2b}{p} + \frac{\tilde{l}}{q} - M \right) c_n^q \\ &< 0. \end{aligned}$$

Secondly, when $B^{0^+} < +\infty$, let $\lambda \in \left(\frac{2bq+2aq+p\tilde{l}}{pqB^{0^+}}, \frac{4^p + l_*(a+b+2)^{p-1}}{p(a+b+2)^{p-1} A_{0^+}} \right)$, choose $\varepsilon_0 > 0$ such that

$$\frac{2a+2b}{p} + \frac{\tilde{l}}{q} - \lambda(B^{0^+} - \varepsilon_0) < 0.$$

Then, there is a positive sequence $\{c_n\} \subset (0, \delta)$ such that $\lim_{n\to+\infty} c_n = 0$ and

$$(B^{0^+} - \varepsilon_0) c_n^q \leq \sum_{t=1}^{b} \sum_{s=1}^{a} F^+((s,t), c_n) = \sum_{t=1}^{b} \sum_{s=1}^{a} F((s,t), c_n) \leq (B^{0^+} + \varepsilon_0) c_n^q.$$

By the definition of the sequence $\{\eta_n\}$ in Y being the same as the case where $B^{0^+} = +\infty$, we have the following:

$$
\begin{aligned}
I_\lambda^+(\eta_n) &\leq \left(\frac{2a+2b}{p}\right)c_n^p + \frac{\tilde{l}}{q}c_n^q - \lambda\left(\sum_{t=1}^{b}\sum_{s=1}^{a} F((s,t),c_n)\right) \\
&\leq \left(\frac{2a+2b}{p}\right)c_n^p + \frac{\tilde{l}}{q}c_n^q - \lambda(B^{0^+} - \varepsilon_0)c_n^q \\
&\leq \left(\frac{2a+2b}{p} + \frac{\tilde{l}}{q} - \lambda(B^{0^+} - \varepsilon_0)\right)c_n^q \\
&< 0.
\end{aligned}
$$

According to the above discussion, we have $I_\lambda^+(\eta_n) < 0$.

Since $I_\lambda^+(0,0,\cdots,0) = 0$ and $\lim_{n\to\infty} \eta_n = (0,0,\cdots,0)$. By combining the above two cases, we obtain that $(0,0,\cdots,0) \in Y$ is a global minima of Φ but $(0,0,\cdots,0)$ is not a local minima of I_λ^+.

Through the above discussion, I_λ^+ satisfies every condition of Lemma 2. According to Lemma 2, there exists a sequence $\{u_n\}$ of pairwise distinct critical points (local minima) of I_λ^+ such that $\lim_{n\to+\infty} u_n = 0$. So, $(s,t) \in \mathbf{Z}(1,a) \times \mathbf{Z}(1,b)$, $y(s,t)$ is a non-zero solution of problem (D^{λ,μ^+}), by Lemma 1, $y(s,t)$ is a positive solution of problem $(D^{\lambda,\mu})$. Therefore, the proof of Theorem 1 is completed. □

Remark 1. *When the nonlinear terms f and g are symmetric on the origin, i.e., $f(\cdot,-y) = -f(\cdot,y)$, $g(\cdot,-y) = -g(\cdot,y)$, it is easy to obtain infinitely many small solutions to problem $(D^{\lambda,\mu})$ by using the critical point theory with symmetries. However, in this paper, we obtain infinitely many small solutions to problem $(D^{\lambda,\mu})$ without the symmetry on f.*

When $\lambda = 1$, according to Theorem 1, we obtain the following.

Corollary 2. *Let $f((s,t),y)$ is a continuous function of y, and $f((s,t),0) \geq 0$, $g((s,t),\cdot) \in C(\mathbf{R},\mathbf{R})$ for every $(s,t) \in \mathbf{Z}(1,a) \times \mathbf{Z}(1,b)$. Suppose the following:*

(i_2) $\frac{p(a+b+2)^{p-1}A_{0^+}}{4^p + l_*(a+b+2)^{p-1}} < 1 < \frac{pqB^{0^+}}{2bq+2aq+p\tilde{l}}$,

(g_1) *there exsits $\delta > 0$ such that at $[0,\delta]$, $G((s,t),y) \geq 0$ and $C^{0^+} < +\infty$.*

Then, for each $\mu \in [0,\bar{\mu}_1^+)$, the following problem (D^μ)

$-\left[\Delta_1(\phi_p(\Delta_1 y(s-1,t))) + \Delta_2(\phi_p(\Delta_2 y(s,t-1)))\right] + l(s,t)\phi_q(y(s,t)) = f((s,t), y(s,t)) + \mu g((s,t),y(s,t))$, $(s,t) \in \mathbf{Z}(1,a) \times \mathbf{Z}(1,b)$, *with Dirichlet boundary conditions (1), has a sequence of positive solutions which converges to zero.*

Similarly, we obtain the following results.

Theorem 2. *Let $f((s,t),y)$ is a continuous function of y, and $f((s,t),0) \leq 0$, $g((s,t),\cdot) \in C(\mathbf{R},\mathbf{R})$ for every $(s,t) \in \mathbf{Z}(1,a) \times \mathbf{Z}(1,b)$. Suppose the following:*

(i_3) $\frac{p(a+b+2)^{p-1}A_{0_-}}{4^p + l_*(a+b+2)^{p-1}} < \frac{pqB^{0_-}}{2bq+2aq+p\tilde{l}}$,

(g_2) *there exsits $\delta > 0$ such that at $[-\delta,0]$, $G((s,t),y) \geq 0$ and $C^{0_-} < +\infty$.*

Then, for every $\lambda \in \left(\frac{2bq+2aq+p\tilde{l}}{pqB^{0_-}}, \frac{4^p + l_*(a+b+2)^{p-1}}{p(a+b+2)^{p-1}A_{0_-}}\right)$ and $\mu \in [0,\bar{\mu}_\lambda^-)$, problem $(D^{\lambda,\mu})$ has a sequence of negative solutions which converges to zero.

When $\lambda = 1$, according to Theorem 2, we obtain the following.

Corollary 3. *Let $f((s,t),y)$ is a continuous function of y, and $f((s,t),0) \leq 0$, $g((s,t),\cdot) \in C(\mathbf{R},\mathbf{R})$ for every $(s,t) \in \mathbf{Z}(1,a) \times \mathbf{Z}(1,b)$. Suppose the following:*

(i_4) $\frac{p(a+b+2)^{p-1}A_{0-}}{4^p+l_*(a+b+2)^{p-1}} < 1 < \frac{pqB^{0-}}{2bq+2aq+p\tilde{l}}$,
(g_2) *there exsits* $\delta > 0$ *such that at* $[-\delta, 0]$, $G((s,t),y) \geq 0$ *and* $C^{0-} < +\infty$.
Then, for each $\mu \in [0, \bar{\mu}_1^-)$, *problem* (D^{μ}) *has a sequence of negative solutions, which converges to zero.*

Combining Theorem 1 with Theorem 2, we have the following.

Theorem 3. *Let* $f((s,t),y)$ *is a continuous function of* y, *and* $f((s,t),0) = 0$, $g((s,t),\cdot) \in C(\mathbf{R},\mathbf{R})$ *for every* $(s,t) \in \mathbf{Z}(1,a) \times \mathbf{Z}(1,b)$. *Suppose that*
(i_5) $\frac{p(a+b+2)^{p-1}\max\{A_{0+},A_{0-}\}}{4^p+l_*(a+b+2)^{p-1}} < \frac{pq\min\{B^{0^+},B^{0-}\}}{2bq+2aq+p\tilde{l}}$,
(g_3) *there exists* $\delta > 0$ *such that at* $[-\delta, \delta]$, $G((s,t),y) \geq 0$ *and* $C^{0*} < +\infty$.
Then, for every $\lambda \in \left(\frac{2bq+2aq+p\tilde{l}}{pq\min\{B^{0^+},B^{0-}\}}, \frac{4^p+l_*(a+b+2)^{p-1}}{p(a+b+2)^{p-1}\max\{A_{0+},A_{0-}\}}\right)$ *and* $\mu \in [0, \min\{\bar{\mu}_\lambda^+, \bar{\mu}_\lambda^-\})$, *problem* $(D^{\lambda,\mu})$ *has two sequences of constant-sign solutions, which converge to zero (with one being positive and the other being negative).*

When $\lambda = 1$, according to Theorem 3, we acquire the following.

Corollary 4. *Let* $f((s,t),y)$ *is a continuous function of* y, *and* $f((s,t),0) = 0$, $g((s,t),\cdot) \in C(\mathbf{R},\mathbf{R})$ *for every* $(s,t) \in \mathbf{Z}(1,a) \times \mathbf{Z}(1,b)$. *Suppose the following:*
(i_6) $\frac{p(a+b+2)^{p-1}A_*}{4^p+l_*(a+b+2)^{p-1}} < 1 < \frac{pqB^{0*}}{2bq+2aq+p\tilde{l}}$,
(g_3) *there exsits* $\delta > 0$ *such that at* $[-\delta, \delta]$, $G((s,t),y) \geq 0$ *and* $C^{0*} < +\infty$.
Then, for each $\mu \in [0, \min\{\bar{\mu}_1^+, \bar{\mu}_1^-\})$, *problem* (D^{μ}) *has two sequences of constant-sign solutions which converge to zero (with one being positive and the other being negative).*

Remark 2. *As a special case of Theorem 1, when* $\mu = 0$.
Considering the following problem, namely (D^{λ})
$-[\Delta_1(\phi_p(\Delta_1 y(s-1,t))) + \Delta_2(\phi_p(\Delta_2 y(s,t-1)))] + l(s,t)\phi_q(y(s,t)) = \lambda f((s,t), y(s,t))$, $(s,t) \in \mathbf{Z}(1,a) \times \mathbf{Z}(1,b)$, *with Dirichlet boundary conditions (1).*

Theorem 4. *Let* $f((s,t),y)$ *be a continuous function of* y, *and* $f((s,t),0) \geq 0$ *for every* $(s,t) \in \mathbf{Z}(1,a) \times \mathbf{Z}(1,b)$. *Suppose the following:*
(i_7) $\frac{p(a+b+2)^{p-1}A_{0+}}{4^p+l_*(a+b+2)^{p-1}} < \frac{pqB^{0^+}}{2bq+2aq+p\tilde{l}}$.
Then, for each $\lambda \in \left(\frac{2bq+2aq+p\tilde{l}}{pqB^{0^+}}, \frac{4^p+l_*(a+b+2)^{p-1}}{p(a+b+2)^{p-1}A_{0+}}\right)$, *problem* (D^{λ}) *has a sequence of positive solutions, which converges to zero.*

4. Example

We provide an example to illustrate our Theorem 3.

Example 1. *Suppose that* $l(s,t) = s + t$. *Let* $a = 2$, $b = 2$, $p = 3$, $q = 2$, *and* $\tilde{l} = \sum_{t=1}^{2}\sum_{s=1}^{2} l(s,t) = 12 < \frac{16+18l_*}{3} = \frac{52}{3}$, *f and g are two functions defined as follows:*

$$f((s,t),c) = f(c) = \begin{cases} \frac{5}{4}pc^{p-1} + pc^{p-1}\sin(\frac{1}{5}\ln c^p) + \frac{1}{5}pc^{p-1}\cos(\frac{1}{5}\ln c^p), & c > 0, \\ 0, & c \leq 0, \end{cases} \tag{20}$$

and

$$g((s,t),c) = g(c) = 2pc^{p-1}. \tag{21}$$

Then, for each $\lambda_1 \in (\frac{49}{54}, \frac{34}{27})$ *and* $\mu_1 \in [0, (\frac{34}{216} - \frac{1}{8}\lambda_1))$, *the following problem, namely* (D^{λ_1,μ_1}).

$-[\Delta_1(\phi_p(\Delta_1 y(s-1,t))) + \Delta_2(\phi_p(\Delta_2 y(s,t-1)))] + l(s,t)\phi_q(y(s,t)) = \lambda_1 f((s,t), y(s,t)) + \mu_1 g((s,t), y(s,t)), (s,t) \in \mathbf{Z}(1,2) \times \mathbf{Z}(1,2)$, *with the following Dirichlet boundary conditions:*

$$y(s,0) = y(s,2) = 0, \quad s \in \mathbf{Z}(0,3),$$
$$y(0,t) = y(2,t) = 0, \quad t \in \mathbf{Z}(0,3),$$

possesses two sequences of constant-sign solutions, which converge to zero (with one being positive and the other being negative).

In fact,

$$F((s,t),c) = \int_0^c f((s,t),\tau)d\tau = \begin{cases} \frac{5}{4}c^p + c^p \sin(\frac{1}{5}\ln c^p), & c > 0, \\ 0, & c \le 0, \end{cases} \tag{22}$$

$$G((s,t),c) = \int_0^c g((s,t),\tau)d\tau = 2c^p. \tag{23}$$

Since $f((s,t),c) > 0, g((s,t),c) > 0$ for $c > 0$, we know that $F((s,t),c)$ and $G((s,t),c)$ are increasing at $c \in (0,+\infty)$. Thus, $\max_{0 \le m \le c} F((s,t),m) = F((s,t),c)$ and $\max_{0 \le m \le c} G((s,t),m) = G((s,t),c)$, for every $c > 0$. Obviously,

$$A_{0*} = \liminf_{c \to 0^+} \frac{abF((s,t),c)}{c^p} = \liminf_{c \to 0^+} \frac{4(\frac{5}{4}c^p + c^p \sin(\frac{1}{5}\ln c^p))}{c^p} = 1,$$

$$B^{0*} = \limsup_{c \to 0^+} \frac{abF((s,t),c)}{c^p} = \limsup_{c \to 0^+} \frac{4(\frac{5}{4}c^p + c^p \sin(\frac{1}{5}\ln c^p))}{c^p} = 9.$$

We can verify condition (i_5) of Theorem 3 since

$$\frac{p(a+b+2)^{p-1}\max\{A_{0^+}, A_{0_-}\}}{4^p + l_*(a+b+2)^{p-1}} = \frac{27}{34} < \frac{pq\min\{B^{0^+}, B^{0_-}\}}{2bq + 2aq + p\tilde{l}} = \frac{54}{49}.$$

Next, we can further verify condition (g_3) of Theorem 3 since

$$C^{0*} = \limsup_{c \to 0^+} \frac{\sum_{t=1}^{b}\sum_{s=1}^{a} G((s,t),c)}{c^p} = \limsup_{c \to 0^+} \frac{ab2c^p}{c^p} = 8 < +\infty.$$

In summary, every condition of Theorem 3 is met.

Therefore, for each $\lambda_1 \in (\frac{49}{54}, \frac{34}{27})$ and $\mu_1 \in [0, (\frac{34}{216} - \frac{1}{8}\lambda_1))$, problem (D^{λ_1,μ_1}) possesses two sequences of constant-sign solutions, which converge to zero (with one being positive and the other being negative).

5. Conclusions

In this paper, we studied the existence of small solutions of perturbed partial discrete Dirichlet problems with the (p,q)-Laplacian operator. Unlike the results in [25], we obtained some sufficient conditions of the existence of infinitely many small solutions, as shown in Theorems 1–3. Firstly, according to Theorem 4.3 of [28] and Lemma 1 of this paper, we obtained a sequence of positive solutions, which converges to zero in Theorem 1. Furthermore, by truncation techniques, we acquired two sequences of constant-sign solutions, which converge to zero (with one being positive and the other being negative). Secondly, the Corollaries 2–4 was acquired when $\lambda = 1$. Finally, as a special case of Theorem 1, we obtained a sequence of positive solutions, which converges to zero in Theorem 4. The existence of large constant-sign solutions of partial difference equations with the (p,q)-Laplacian operator will be discussed by the method used in this paper as our future research direction.

Author Contributions: All authors contributed equally to this paper. All authors have read and agreed to the published version of the manuscript.

Funding: This work is supported by the National Natural Science Foundation of China (grant No. 11971126), the Program for Changjiang Scholars and Innovative Research Team in University (grant No. IRT_16R16) and the Innovation Research for the Postgraduates of Guangzhou University (grant No. 2020GDJC-D07).

Institutional Review Board Statement: Not applicable.

Informed Consent Statement: Not applicable.

Data Availability Statement: Not applicable.

Conflicts of Interest: The authors declare no conflict of interests.

References

1. Elaydi, S. *An Introduction to Difference Equations*, 3rd ed.; Springer: New York, NY, USA, 2005.
2. Long, Y.; Wang, L. Global dynamics of a delayed two-patch discrete SIR disease model. *Commun. Nonlinear Sci. Numer. Simul.* **2020**, *83*, 105117. [CrossRef]
3. Yu, J.; Zheng, B. Modeling Wolbachia infection in mosquito population via discrete dynamical model. *J. Differ. Equ. Appl.* **2019**, *25*, 1549–1567. [CrossRef]
4. Agarwal, R. *Difference Equations and Inequalities: Theory, Methods, and Applications*; Marcel Dekker: New York, NY, USA, 1992.
5. Henderson, J.; Thompson, H. Existence of multiple solutions for second order discrete boundary value problems. *Comput. Math. Appl.* 2002, 43, 1239–1248. [CrossRef]
6. Bereanu, C.; Mawhin, J. Boundary value problems for second-order nonlinear difference equations with discrete ϕ-Laplacian and singular ϕ. *J. Differ. Equ. Appl.* **2008**, *14*, 1099–1118. [CrossRef]
7. Wang, S.; Long, Y. Multiple solutions of fourth-order functional difference equation with periodic boundary conditions. *Appl. Math. Lett.* **2020**, *104*, 106292. [CrossRef]
8. Long, Y. Existence of multiple and sign-changing solutions for a second-order nonlinear functional difference equation with periodic coefficients. *J. Differ. Equ. Appl.* **2020**, *26*, 966–986. [CrossRef]
9. Long, Y.; Chen, J. Existence of multiple solutions to second-order discrete Neumann boundary value problem. *Appl. Math. Lett.* **2018**, *83*, 7–14. [CrossRef]
10. Guo, Z.; Yu, J. The existence of periodic and subharmonic solutions for second-order superlinear difference equations. *Sci. China Ser. A Math.* **2003**, *46*, 506–515. [CrossRef]
11. Shi, H. Periodic and subharmonic solutions for second-order nonlinear difference equations. *J. Appl. Math. Comput.* **2015**, *48*, 157–171. [CrossRef]
12. Zhou, Z.; Ma, D. Multiplicity results of breathers for the discrete nonlinear Schrödinger equations with unbounded potentials. *Sci. China Math.* **2015**, *58*, 781–790. [CrossRef]
13. Zhou, Z.; Yu, J.; Chen, Y. Homoclinic solutions in periodic difference equations with saturable nonlinearity. *Sci. China Math.* **2011**, *54*, 83–93. [CrossRef]
14. Lin, G.; Zhou, Z. Homoclinic solutions of discrete ϕ-Laplacian equations with mixed nonlinearities. *Commun. Pure Appl. Anal.* **2018**, *17*, 1723–1747. [CrossRef]
15. Zhang, Q. Homoclinic orbits for a class of discrete periodic Hamiltonian systems. *Proc. Am. Math. Soc.* **2015**, *143*, 3155–3163. [CrossRef]
16. Zhang, Q. Homoclinic orbits for discrete Hamiltonian systems with indefinite linear part. *Commun. Pure Appl. Anal.* **2017**, *14*, 1929–1940. [CrossRef]
17. Zhang, Q. Homoclinic orbits for discrete Hamiltonian systems with local super-quadratic conditions. *Commun. Pure Appl. Anal.* **2019**, *18*, 425–434. [CrossRef]
18. Lin, G.; Zhou, Z.; Yu, J. Ground state solutions of discrete asymptotically linear Schrödinger equations with bounded and non-periodic potentials. *J. Dyn. Differ. Equ.* **2020**, *32*, 527–555. [CrossRef]
19. Bonanno, G.; Candito, P. Infinitely many solutions for a class of discrete non-linear boundary value problems. *Appl. Anal.* **2009**, *88*, 605–616. [CrossRef]
20. Bonanno, G.; Jebelean, P.; Serban, C. Superlinear discrete problems. *Appl. Math. Lett.* **2016**, *52*, 162–168. [CrossRef]
21. D'Agua, G.; Mawhin, J.; Sciammetta, A. Positive solutions for a discrete two point nonlinear boundary value problem with p-Laplacian. *J. Math. Anal. Appl.* **2017**, *447*, 383–397.
22. Nastasi, A.; Vetro, C.; Vetro, F. Positive solutions of discrete boundary value problems with the (p,q)-Laplacian operator. *Electron. J. Differ. Equ.* **2017**, *225*, 1–12.
23. Zhou, Z.; Ling, J. Infinitely many positive solutions for a discrete two point nonlinear boundary value problem with ϕ_c-Laplacian. *Appl. Math. Lett.* **2019**, *91*, 28–34. [CrossRef]

24. Ling, J.; Zhou, Z. Positive solutions of the discrete Dirichlet problem involving the mean curvature operator. *Open Math.* **2019**, *17*, 1055–1064. [CrossRef]
25. Wang, J.; Zhou, Z. Large constant-sign solutions of discrete Dirichlet boundary value problems with p-mean curvature operator. *Mathematics* **2020**, *8*, 381. [CrossRef]
26. Heidarkhani, S.; Imbesi, M. Multiple solutions for partial discrete Dirichlet problems depending on a real parameter. *J. Differ. Equ. Appl.* **2015**, *21*, 96–110. [CrossRef]
27. Du, S.; Zhou, Z. Multiple solutions for partial discrete Dirichlet problems involving the p-Laplacian. *Mathematics* **2020**, *8*, 2030. [CrossRef]
28. Bonanno, G.; Candito, P.; D'Agui, G. Variational methods on finite dimensional Banach spaces and discrete problems. *Adv. Nonlinear Stud.* **2014**, *14*, 915–939. [CrossRef]

symmetry

MDPI

Article

Vibration Properties of a Concrete Structure with Symmetries Used in Civil Engineering

Sorin Vlase [1,2], Marin Marin [3,*] and Ovidiu Deaconu [4]

1 Department of Mechanical Engineering, Transilvania University of Brașov, B-dul Eroilor 20, 500036 Brasov, Romania; svlase@unitbv.ro
2 Romanian Academy of Technical Sciences, B-dul Dacia 26, 030167 Bucharest, Romania
3 Department of Mathematics and Computer Science, Transilvania University of Brasov, B-dul Eroilor 29, 500036 Brasov, Romania
4 Department of Civil Engineering, Transilvania University of Brasov, 500036 Brasov, Romania; ovideaconu@unitbv.ro
* Correspondence: m.marin@unitbv.ro; Tel.: +40-268-722643020

Abstract: The paper aims to study a concrete structure, currently used in civil engineering, which has certain symmetries. This type of problem is common in engineering practice, especially in civil engineering. There are many reasons why structures with identical elements or certain symmetries are used in industry, related to economic considerations, shortening the design time, for constructive, simplicity, cost or logistical reasons. There are many reasons why the presence of symmetries has benefits for designers, builders, and beneficiaries. In the end, the result of these benefits materializes through short execution times and reduced costs. The paper studies the eigenvalue and eigenmode properties of vibration for components of the constructions' structure, often encountered in current practice. The identification of such properties allows the simplification and easing of the effort necessary for the dynamic analysis of such a structure.

Keywords: vibrations; symmetrical structures; eigenmodes; building; concrete

Citation: Vlase, S.; Marin, M.; Deaconu, O. Vibration Properties of a Concrete Structure with Symmetries Used in Civil Engineering. *Symmetry* **2021**, *13*, 656. https://doi.org/10.3390/sym13040656

Academic Editors: Jan Awrejcewicz and Calogero Vetro

Received: 22 March 2021
Accepted: 9 April 2021
Published: 12 April 2021

1. Introduction

Frequently encountered in the design and construction of structures used in civil engineering, symmetries allow, in many cases, the simplification of calculations and dynamic analysis of such a structure. The direct consequence would be the shortening of the design and execution time and, of course, the decrease of the costs generated by these stages. The smaller information provided by a repetitive or symmetrical structure can help ease the computational effort. In the case of a static calculation, methods of approaching this problem are presented in the Strength of Materials courses. In the dynamic case, considering the elastic elements and studying the vibrations, although certain properties have long been observed [1], a systematic study of the problem has not yet been done. A case in which the symmetries introduced by two identical motors and their effect on vibrations were considered was studied [2]. The use of identical systems was applied to the design of a centrifugal pendulum vibration absorber system [3]. Circular symmetry and its induced properties have been reported previously [4,5]. Other particular cases have been studied [6] using a finite difference method and [7] for continuous systems. In the following, we will study the case of a mechanical system consisting of four trusses, two of which are identical. The transverse and torsional vibrations of such a system that are strongly coupled for the chosen case will be studied.

In the field of engineering, not only in civil engineering, but also in other fields such as the machine or machinery manufacturing industry, the automotive industry, and the aerospace industry, there are products, parts of products, machines, and components that contain identical, repetitive elements, which have, in their composition, parts that show symmetries of different types.

Until now, symmetries in Mechanics have been studied mainly from the point of view of the mathematics involved [8], as they have effects in writing equations of motion [9], but their applications in practice are little studied [10]. In January 2018, a special issue of the Symmetry review dedicated to applications in Structural Mechanics (Civil Engineering and Symmetry-2018, ISSN 2073-8994, [11]) was launched. The course Similarity, Symmetry and Group Theoretical Methods in Mechanics was organized at the Center for Solid Mechanics-CISM from UDINE (see ref. [12]).

The use of substructures in the design of the aeronautical industry is a relatively common procedure used to reduce working time. Finite element models of different parts are condensed into substructures. Obviously a substructure contains less degree of freedom (DOF) than the entire structure and it is easier to model it. Likewise, for identical parts that are found in the construction and design of aeronautical panels, if the substructure has been generated for one part, the generation of the whole assembly becomes easier to do. A method reduces the size of the model, useful in the design stage and in the manufacture of the entire structure. The accuracy of the finite element analysis performed is maintained. An illustration of this method is presented in a previous work [13].

The modeling of mechanical systems with repetitive or identical parts leads, finally, to systems of differential equations that describe the answer of such systems, which have, in their component, strings of identical terms. This feature leads to simpler methods of solving these systems of differential equations. Such an issue is addressed in studies [14,15].

The vast majority of buildings, works of art, halls, and in general, the constructions, have identical parts and have symmetries. It is a situation that has existed since the beginning of the first constructions made by man (antiquity) and the reasons are of several kinds as an easier, faster design, then a cheaper realization and, less important for engineers but important for beneficiaries, for aesthetic reasons. The structures have in their composition repetitive elements or present different forms of symmetry. These properties can be used successfully to facilitate static and dynamic analysis.

Group theory has been used extensively to study various phenomena in physics and chemistry, such as quantum mechanics, crystallography, and molecular structure. However, this theory can find a fertile field of application in engineering. This allows simplifying the analysis of systems that have certain symmetries or identical parts. In this way, it was efficient in the study of vibrations or the dynamic or kinematic analysis of mechanical systems. The use of group theory in engineering was analyzed [16,17]. Different aspects of the use of symmetry in engineering are presented in other papers. In a research [18], the influence of the symmetry of boundary condition in the description of the models was studied. Some theoretical basis and an attempt to classify the symmetries that occur in structures were presented [19] and application of the symmetry in engineering structures were presented [20,21].

In recent years, new and interesting methods of studying this type of problem have been studied by researchers [22,23] and new ways to deal with such problems were reported [24,25].

However, there are still many situations that can be studied and, therefore, the present paper aims to complete the cases studied and to offer proposals for the application of these properties that could help a design engineer to ease his effort.

2. Model and Free Vibration Response

We have 2 coplanar reinforced concrete beams with different properties. AD beam is a main beam, considered double clamped at both ends, with the length $L_1 + L_2$. CB beam is a secondary beam, considered simply supported at both ends (nodes C and B), with length $L_3 + L_3$. The beams are made monolithically, so at point O of the intersection, a rigid knot is created. The secondary beam is arranged symmetrically to the main beam. The 2 beams have the Young's moduli E_1 and E_2. The 2 beams have different sectional properties: for beam AD, we have the moment of inertia Iz1 and the area of section A_1, and for beam

CB, we have the moment of inertia I_{z2} and the area of section A_2, with the property that $I_{z1} > I_{z2}$ and $A_1 > A_2$. The whole structure is presented in Figure 1.

Figure 1. Structure with repetitive cells.

Thus, the model of the mechanical system considered (Figure 2) consists of two identical trusses OB and OC rigidly fixed, perpendicular to a third bar AOD. The trusses can have transverse vibrations in a direction perpendicular to the ABDC plane and torsional vibrations. The trusses are clamped in the points B and C. In A and D, the AOD trusses are clamped, so the displacement, slope, and torsion angle at these points are zero. For point O, the transverse displacements of point O belonging to all four bars are equal. The torsion angle of the truss AO in O is equal to the torsion angle of the truss OD in O and with the slope of the bars OB and OC in O. The torsion angle of the trusses OB and OC in O is equal to the slope of the trusses AO and OD in point O. The sum of the shear forces and moments in O will be zero.

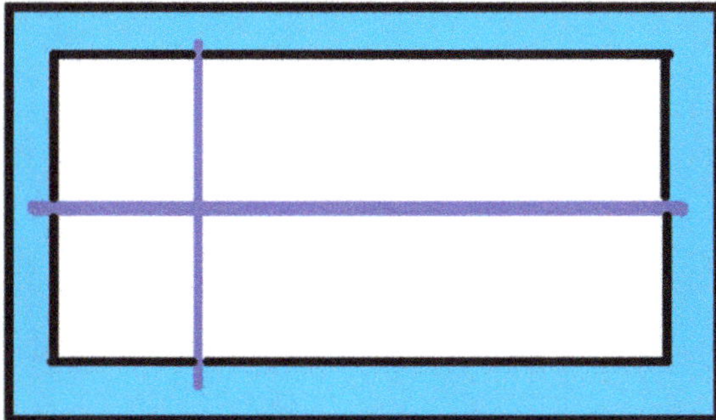

Figure 2. One repetitive cell.

We will study a continuous concrete truss, homogeneous, with constant section. If there are no forces distributed or concentrated along the length of the truss, the vibrations of this are described by the well-known Equations [26]:

$$\frac{\partial^4 v}{\partial x^4} + \frac{\rho A}{E I_z} \frac{\partial^2 v}{\partial t^2} = 0 \tag{1}$$

The notations used in Equation (1) are the following: v—is the truss deflection, A—s the cross section, ρ is the mass density, E—Young's modulus, and I_z is the second area moment of inertia with respect to the z axis and x is the ordinate of the point having the deflection v.

To solve Equation (1), we look for a solution of the form [27–29]:

$$v(x,t) = \Phi(x)\sin(pt + \theta). \tag{2}$$

Equation (2) must check Equation (1) at any time and, imposing this condition, we obtain:

$$\frac{\partial^4\Phi}{\partial x^4} - p^2\frac{\rho A}{EI_z}\Phi = 0 \tag{3}$$

Denote:

$$\lambda^4 = \frac{\rho A}{EI_z} \tag{4}$$

Then, Equation (1) becomes:

$$\frac{\partial^4\Phi}{\partial x^4} - p^2\lambda^4\Phi = 0 \tag{5}$$

In Equation (5), Φ represents the function, depending on the abscissa x, which gives us the deformed bar (eigenmode) corresponding to its eigenpulsation p. The solution is:

$$\Phi(x) = C_1\sin(\lambda\sqrt{p}x) + C_2\cos(\lambda\sqrt{p}x) + C_3\mathrm{sh}(\lambda\sqrt{p}x) + C_4\mathrm{ch}(\lambda\sqrt{p}x). \tag{6}$$

The constants C_1, C_2, C_3, C_4 are determined considering the boundary conditions for this problem.

In the following, we will use Equation (5) for the domains defined by the four trusses, obtaining, in this way, four differential equations of the fourth order, corresponding to the frames AO, OD, OB, OC (see Appendix A).

The study of torsional vibrations for a straight bar, unloaded over the length, leads to the second order differential equation:

$$\frac{\partial^2\varphi}{\partial x^2} - \frac{J}{GI_p}\frac{\partial^2\varphi}{\partial t^2} = 0 \tag{7}$$

where φ is the angle of torsion of the cross section being at the distance x from the end of the truss, $J = \rho I_p$ is the polar moment of inertia, I_p is the polar second moment of the area, and G is shear's modulus.

The solution Equation (7) is sought in the form:

$$\varphi(x,t) = \psi(x)\sin(pt + \theta) \tag{8}$$

Equation (8) must verify Equation (7), from which we obtain:

$$\frac{\partial^2\psi}{\partial x^2} + p^2\delta_i^2\Phi = 0 \tag{9}$$

where the notation was made:

$$\delta^2 = \frac{J}{G\,I_p} \tag{10}$$

The solution is:

$$\psi(x) = D_1\sin(\delta p x) + D_2\cos(\delta p x) \tag{11}$$

Denoted by M^b is the bending moment of a bar in section x, T the shear force that appears in the cross section and with M^t the torque. For the studied system in the paper, the boundary conditions are:

(a) For the AO truss, the end A is clamped, so: $v_{AO}(0,t) = 0$; $v'_{AO}(0,t) = 0$; $\varphi(0,t) = 0$;
(b) For the OD truss, the end D is supported, so: $v_{OD}(L_1,t) = 0$; $v'_{OD}(L_1,t) = 0$; $\varphi(0,t) = 0$;
(c) For the OB truss, the end B is supported so: $v_{OB}(L_3,t) = 0$; $M^b_{OB}(L_3,t) = 0$; $M^t_{OB}(L_3,t) = 0$;

(d) For the OC truss, the end C is clamped, so: $v_{OC}(L_3, t) = 0$; $M^b_{OC}(L_3, t) = 0$; $M^t_{OC}(L_3, t) = 0$.

Imposing these conditions, for the considered four bars, 12 boundary conditions are obtained. The two moments and the shear force are expressed by the known relationships in the mechanics of the deformable solid [30–33]:

$$M^b(x) = -EI_z \frac{\partial^2 v(x)}{\partial x^2};\ T(x) = EI_z \frac{\partial^3 v(x)}{\partial x^3};\ M^t(x) = GI_p \frac{\partial \varphi(x)}{\partial x} \tag{12}$$

Using the Equation (12), the solutions presented in Appendixes, and the boundary conditions (a), (b), (c), and (d) result in 12 linear equations involving the unknown constants (Appendix A).

The continuity of the elastic system at point O leads to the following conditions: we have once that the displacements in C for all trusses: $v_{AO}(L_2, t) = v_{OD}(0, t) = v_{OB}(0, t) = v_{OC}(0, t)$, (three conditions), the slopes in O of the trusses AO and OD are equal to the torsion angle of the trusses OB and OC in: $v'_{AO}(L_2, t) = v'_{OD}(0, t) = \varphi_{OB}(0, t) = -\varphi_{OC}(0, t)$, (three conditions), and the torsion angles of the bars AO and OD in O are equal to the slopes of the bar OB and OC in O: $\varphi_{AO}(L_2, t) = \varphi_{OD}(0, t) = v'_{OB}(0, t) = -v'_{OC}(0, t)$) (three conditions). These mean nine conditions presented in the Appendix B, from which nine linear equations result.

Three more conditions are needed to obtain the constants in the written differential equations. They are obtained by considering the equilibrium of an infinitesimal element of the mass containing the point O.

The sum of the four shear forces in O must be zero, so:

$$T1 + T2 + T3 - T4 = 0 \tag{13}$$

A similar equilibrium relation is obtained for bending and torsional moments:

$$M^{b1} + M^{b2} + M^{t3} - M^{t4} = 0, \tag{14}$$

$$M^{t1} + M^{t2} + M^{b3} - M^{b4} = 0 \tag{15}$$

Writing these three conditions results in three linear equations (Appendix C). The unknowns are:

$$\left\{B^{AO}\right\} = \left[\begin{array}{cccccc} C_1^{AO} & C_2^{AO} & C_3^{AO} & C_4^{AO} & D_1^{AO} & D_2^{AO} \end{array}\right] \tag{16}$$

$$\left\{B^{OD}\right\} = \left[\begin{array}{cccccc} C_1^{OD} & C_2^{OD} & C_3^{OD} & C_4^{OD} & D_1^{OD} & D_2^{OD} \end{array}\right] \tag{17}$$

$$\left\{B^{OB}\right\} = \left[\begin{array}{cccccc} C_1^{OB} & C_2^{OB} & C_3^{OB} & C_4^{OB} & D_1^{OB} & D_2^{OB} \end{array}\right] \tag{18}$$

$$\left\{B^{OC}\right\} = \left[\begin{array}{cccccc} C_1^{OC} & C_2^{OC} & C_3^{OC} & C_4^{OC} & D_1^{OC} & D_2^{OC} \end{array}\right] \tag{19}$$

or:

$$\{B\} = \left\{\begin{array}{c} \{B^{AO}\} \\ \{B^{OD}\} \\ \{B^{OB}\} \\ \{B^{OC}\} \end{array}\right\} \tag{20}$$

In such way, a homogeneous linear system with 24 equations with 24 unknowns was obtained. In order to have other solutions besides the trivial solution zero, the determinant of the system must be zero. Putting this condition, the obtained eigenfrequencies of the system can be determined from the obtained equation.

The continuous models used in our studies are excellent for a classical analysis of such systems. The boundary conditions, written for these models, ultimately lead to a linear system of homogeneous equations. For this system to have a solution other than the trivial solution, zero, it is necessary that the system determinant be equal to zero. Imposing this condition leads to writing the characteristic equation that will provide, by solving, its eigenfrequencies of vibration. Once these values are known, then, we can determine your eigenmodes for these frequencies.

The matrix of the system can be written, in a concise form:

$$A_{24\times 24}\begin{bmatrix} A_{11} & 0 & A_{12} \\ 0 & A_{11} & A_{12} \\ A_{21} & A_{21} & A_{22} \end{bmatrix} \tag{21}$$

where the matrices $[A_{ij}]$ are presented in Appendix C.

$$[A]\{B\} = \{0\} \tag{22}$$

The condition:

$$\det(A) = 0 \tag{23}$$

offers the eigenvalues of the system of differential equations.

3. Properties of the Eigenvalues and Eigenmodes

Let us now consider one of the identical trusses OB and OC. The truss is clamped in O and supported in B (or C). Equations (1) and (7) are also valid for the OB bar (OC), with the boundary conditions:

For point O:

$$x = 0;\ v(0,t) = 0;\ v'(0,t) = 0;\ \varphi(0,t) = 0 \tag{24}$$

and for the point B(C):

$$x = L_3;\ v(L_3,t) = 0;\ M^b(L_3,t) = 0;\ M^t(L_3,t) = 0 \tag{25}$$

The solution is:

$$\Phi(x) = C_1\sin(\lambda_2\sqrt{p}x) + C_2\cos(\lambda_2\sqrt{p}x) + C_3\mathrm{sh}(\lambda_2\sqrt{p}x) + C_4\mathrm{ch}(\lambda_2\sqrt{p}x) \tag{26}$$

for transversal vibrations and:

$$\psi(x) = D_1\sin(\delta_2 px) + D_2\cos(\delta_2 px) \tag{27}$$

with the imposed boundary conditions for this case:

$$C_2^{OB} + C_4^{OB} = 0;\ C_1^{OB} + C_3^{OB} = 0;\ D_2^{OB} = 0 \tag{28}$$

$$C_1^{BO}\sin(\lambda_2\sqrt{p}L_3) + C_2^{BO}\cos(\lambda_2\sqrt{p}L_3) + C_3^{BO}\mathrm{sh}(\lambda_2\sqrt{p}L_3) + C_4^{BO}\mathrm{ch}(\lambda_2\sqrt{p}L_3) = 0 \tag{29}$$

$$-C_1^{BO}\sin(\lambda_2\sqrt{p}\,L_3) - C_2^{BO}\cos(\lambda_2\sqrt{p}L_3) + C_3^{BO}\mathrm{sh}(\lambda_2\sqrt{p}L_3) + C_4^{BO}\mathrm{ch}(\lambda_2\sqrt{p}L_3) = 0 \tag{30}$$

$$D_1^{BO}\cos(\delta_2 pL_3) - D_2^{BO}\sin(\delta_2 pL_3) = 0 \tag{31}$$

Using conditions expressed by Equations (28)–(31), it is now possible to determine the constants C_1^{BO}, C_2^{BO}, C_3^{BO}, C_4^{BO}, D_1^{BO}, D_2^{BO} from the linear homogenous system:

$$[A_{11}]\left\{B^{OB}\right\} = 0 \tag{32}$$

where $[A_{11}]$ is the matrix determined by Equation (A76).

If the condition of the existence of non-zero solutions is set:

$$\det(A_{11}) = 0 \tag{33}$$

It is now possible to obtain the eigenvalues of the truss OB (or OC).

The following theorems will be proved in the following:

Theorem 1. *The eigenvalues for the OB truss, clamped at one end and supported at the other, are also eigenvalues for the entire mechanical system.*

Proof. We must show that det (A) = 0 implies det (S) = 0. In reference [34], this property is proved in a more general case. It turns out that the property is valid in our case. □

It follows that the eigenvalues of a single truss in Appendix D, clamped at one end and supported at the other, are also eigenvalues of the composed system, clamped in A and D and with the ends B and C supported.

Using the matrix done by Equation (A76), and obtaining the eigenvalues for this matrix, the eigenmodes of deformations are obtained using Equation (A85). The following two theorems will be proved:

Theorem 2. *For eigenvalues that are common to the whole mechanical system (Figure 3) and to the subsystem in Figure A1 (see Theorem 1), eigenvectors are of the form:*

$$\Phi = \begin{Bmatrix} \Phi_1 \\ -\Phi_1 \\ 0 \end{Bmatrix} \tag{34}$$

(the existence of common eigenvalues is proved by theorem T1).

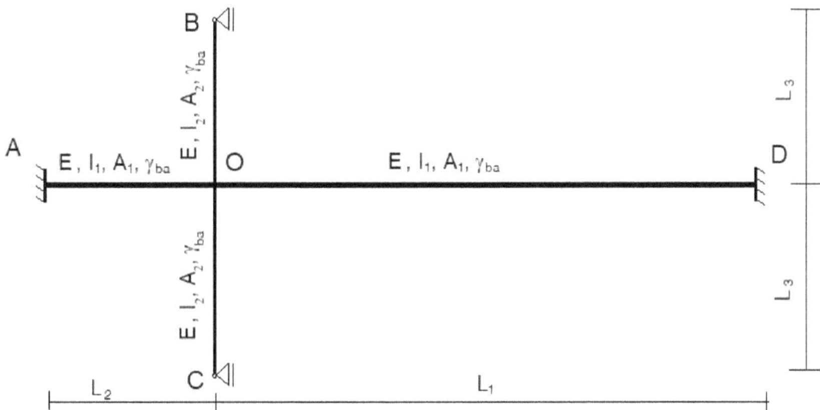

Figure 3. Geometry of a repetitive cell.

Proof. For the eigenvalues obtained from Equation (33), the following system must be solved:

$$\begin{bmatrix} A_{11} & 0 & A_{12} \\ 0 & A_{11} & A_{12} \\ A_{21} & A_{21} & A_{22} \end{bmatrix} \begin{Bmatrix} \Phi_{OB} \\ \Phi_{OC} \\ \Phi_{AOD} \end{Bmatrix} = \begin{Bmatrix} 0 \\ 0 \\ 0 \end{Bmatrix} \tag{35}$$

where:

$$\det A_{11} = 0 \tag{36}$$

Condition (36) implies that a vector Φ_{OB} can be found, such that:

$$A_{11}\Phi_{OB} = 0 \tag{37}$$

Suppose that we determined this vector. Equation (35) becomes, after performing some simple calculations:

$$A_{11}\Phi_{OB} + A_{12}\Phi_{AOD} = 0 \tag{38}$$

$$A_{11}\Phi_{OC} + A_{12}\Phi_{AOD} = 0 \tag{39}$$

$$A_{21}(\Phi_{OB} + \Phi_{OC}) + A_{22}\Phi_{AOD} = 0 \tag{40}$$

If we take into account Equation (37), the system of Equations (38)–(40) becomes:

$$A_{12}\Phi_{AOD} = 0 \tag{41}$$

$$A_{11}\Phi_{OC} + A_{12}\Phi_{AOD} = 0 \tag{42}$$

$$A_{21}(\Phi_{OB} + \Phi_{OC}) + A_{22}\Phi_{AOD} = 0 \tag{43}$$

From Equation (41), because, in general, $\det A_{12} \neq 0$, it follows immediately:

$$\Phi_{AOD} = 0 \tag{44}$$

and introducing that in Equation (43), we obtain $\Phi_{OB} = -\Phi_{OC}$, a relation which verifies also Equation (42), if we take into account Equation (37). If $\Phi_{OB} = \Phi_1$ is denoted, it results in Equation (34). □

Theorem 3. *For the other eigenvalues of the system, the eigenvectors are of the form:*

$$\Phi = \begin{Bmatrix} \Phi_1 \\ \Phi_1 \\ \Phi_3 \end{Bmatrix} \tag{45}$$

Proof. For the eigenvalues calculated, the system of Equation (35) must be solved, with $\det A_{11} \neq 0$ or:

$$A_{11}\Phi_{OB} + A_{12}\Phi_{AOD} = 0 \tag{46}$$

$$A_{11}\Phi_{OC} + A_{12}\Phi_{AOD} = 0 \tag{47}$$

$$A_{21}(\Phi_{OB} + \Phi_{OC}) + A_{22}\Phi_{AOD} = 0 \tag{48}$$

Subtracting (47) from (46), we get:

$$A_{11}(\Phi_{OB} - \Phi_{OC}) = 0 \tag{49}$$

If $\det A \neq 0$, it results in $\Phi_s - \Phi_m = 0$ and $\Phi_s = \Phi_m = \Phi_1$. □

A block diagram representing the stage of the analysis is presented in Figure 4.

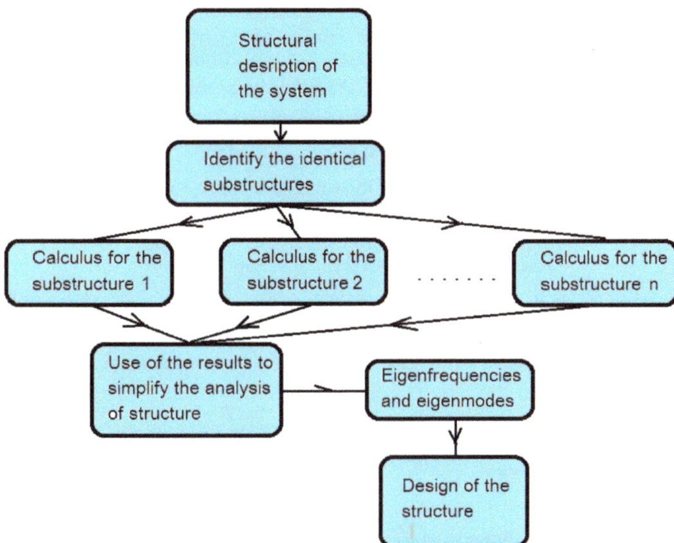

Figure 4. Block diagram of the operation.

For the eigenfrequencies of the system that are the same with the eigenvalues of a single clamped truss at one end and supported at the other, the eigenmodes are skew symmetric, the two identical trusses vibrate in counterphase, and the third truss AOD rests. For the other eigenfrequencies, the identical trusses have identical eigenmodes.

4. Conclusions

Buildings and constructions, in general, show different forms of symmetry or are made up of repetitive elements. The existence of these symmetries leads to obtaining advantages related to the calculation, design, and manufacture of the structure. One of the advantages is related to the ease of describing the system by systematizing the information used; then, the properties demonstrated in the paper allow to decrease the time required to perform calculations and all this will allow savings and simplifications in the design process. Then, the existence of identical elements or identical parts can simplify the process of making the structure, by simplifying the labor and effort required to manufacture the structure. In conclusions, the design is simpler, and the realization costs are lower. There are also aesthetic reasons that justify the realization of structures with symmetries. From the point of view of calculation and behavior in static and dynamic cases, symmetries can bring significant advantages. In the strength of materials, symmetries are widely used in the static analysis of structures. However, they can be used for dynamic analysis, so that the vibrations of such structures allow simplification of the calculation and time savings in the design process. The paper has presented several vibration properties of a symmetrical structure made of concrete, used in civil engineering. Such structures are frequently encountered in the construction of buildings and in civil engineering and the knowledge of vibration properties can prove to be an advantage that allows to reduce the time and costs related to the design. We mention that symmetries appearing in all aspects of current life and in engineering applications are common. In consequence, the results obtained can be extended to other situations that may be encountered in practice. Future research could reveal other types of symmetries that will allow the systematization of the results and the proposal of a strategy to approach the design and execution of systems with identical parts or symmetries.

Author Contributions: Conceptualization, S.V., M.M., and O.D.; methodology, S.V.; validation, S.V., M.M., and O.D.; formal analysis, M.M.; investigation, S.V.; resources, O.D.; writing—original draft preparation, S.V.; writing—review and editing, S.V.; visualization, M.M. and O.D.; supervision, M.M.; project administration, O.D.; funding acquisition, O.D. All authors have read and agreed to the published version of the manuscript.

Funding: This research received no external funding.

Institutional Review Board Statement: Not applicable.

Informed Consent Statement: Not applicable.

Data Availability Statement: Not applicable.

Acknowledgments: We want to thank the reviewers who have read the manuscript carefully and have proposed pertinent corrections that have led to an improvement in our manuscript.

Conflicts of Interest: The authors declare no conflict of interest.

Appendix A

For the truss AO:

$$\frac{\partial^4 \Phi_{AO}}{\partial x^4} - \frac{\rho_1 A_1}{E_1 I_{z1}} p^2 \Phi_{AO} = 0 \tag{A1}$$

For the truss OD:

$$\frac{\partial^4 \Phi_{BC}}{\partial x^4} - \frac{\rho_1 A_1}{E_1 I_{z1}} p^2 \Phi_{OD} = 0 \tag{A2}$$

For the truss OB:

$$\frac{\partial^4 \Phi_{OB}}{\partial x^4} - \frac{\rho_2 A_2}{E_2 I_{z2}} p^2 \Phi_{OB} = 0 \tag{A3}$$

For the truss OC:

$$\frac{\partial^4 \Phi_{OC}}{\partial x^4} - \frac{\rho_2 A_2}{E_2 I_{z2}} p^2 \Phi_{OC} = 0 \tag{A4}$$

The following notations are made:

$$\frac{\rho_1 A_1}{E_1 I_{z1}} = \lambda_1^4;\ \frac{\rho_2 A_2}{E_2 I_{z2}} = \lambda_2^4 \tag{A5}$$

Using (A5), the four solutions for the four differential equations of order four (A1)–(A4) are:

$$\Phi_{AO}(x) = C_1^{AO} \sin(\lambda_1 \sqrt{p} x) + C_2^{AO} \cos(\lambda_1 \sqrt{p} x) + C_3^{AO} \mathrm{sh}(\lambda_1 \sqrt{p} x) + C_4^{AO} \mathrm{ch}(\lambda_1 \sqrt{p} x) \tag{A6}$$

$$\Phi_{OD}(x) = C_1^{OD} \sin(\lambda_1 \sqrt{p} x) + C_2^{OD} \cos(\lambda_1 \sqrt{p} x) + C_3^{OD} \mathrm{sh}(\lambda_1 \sqrt{p} x) + C_4^{OD} \mathrm{ch}(\lambda_1 \sqrt{p} x) \tag{A7}$$

$$\Phi_{OB}(x) = C_1^{OB} \sin(\lambda_2 \sqrt{p} x) + C_2^{OB} \cos(\lambda_2 \sqrt{p} x) + C_3^{OB} \mathrm{sh}(\lambda_2 \sqrt{p} x) + C_4^{OB} \mathrm{ch}(\lambda_2 \sqrt{p} x) \tag{A8}$$

$$\Phi_{OC}(x) = C_1^{OC} \sin(\lambda_2 \sqrt{p} x) + C_2^{OC} \cos(\lambda_2 \sqrt{p} x) + C_3^{OC} \mathrm{sh}(\lambda_2 \sqrt{p} x) + C_4^{OC} \mathrm{ch}(\lambda_2 \sqrt{p} x) \tag{A9}$$

For torsion, the notation was made:

$$\delta_i^2 = \frac{J_i}{G_i I_{pi}}\ i = 1,2 \tag{A10}$$

Index 1 corresponds to trusses AO and OD and index 2 to trusses OB and OC. Applying Equation (8) for the four trusses studied, leads us to:

For torsional vibrations of the bar AO:

$$\frac{\partial^2 \psi_{AO}}{\partial x^2} + p^2 \delta_1^2 \psi_{AO} = 0 \tag{A11}$$

for torsional vibrations of the bar OD:

$$\frac{\partial^2 \psi_{OD}}{\partial x^2} + p^2 \delta_1^2 \psi_{OD} = 0 \tag{A12}$$

for torsional vibrations of the bar OB:

$$\frac{\partial^2 \psi_{OB}}{\partial x^2} + p^2 \delta_2^2 \psi_{OB} = 0 \tag{A13}$$

for torsional vibrations of the bar OC:

$$\frac{\partial^2 \psi_{OC}}{\partial x^2} + p^2 \delta_2^2 \psi_{OC} = 0 \tag{A14}$$

The solutions of the four equations (A11)–(A14) are:

$$\psi_{AO}(x) = D_1^{AO} \sin(\delta_1 p x) + D_2^{AO} \cos(\delta_1 p x) \tag{A15}$$

$$\psi_{OD}(x) = D_1^{OD} \sin(\delta_1 p x) + D_2^{OD} \cos(\delta_1 p x) \tag{A16}$$

$$\psi_{OB}(x) = D_1^{OB} \sin(\delta_2 p x) + D_2^{OB} \cos(\delta_2 p x) \tag{A17}$$

$$\psi_{OC}(x) = D_1^{OC} \sin(\delta_2 p x) + D_2^{OC} \cos(\delta_2 p x) \tag{A18}$$

The solutions contain 24 integration constants that will be determined considering the boundary conditions.

The two moments and the shear force are expressed by the known relationships in the mechanics of the deformable solid [26,27]:

$$M^b(x) = -EI_z \frac{\partial^2 v(x)}{\partial x^2};\ T(x) = EI_z \frac{\partial^3 v(x)}{\partial x^3};\ M^t(x) = GI_p \frac{\partial \varphi(x)}{\partial x}$$

However:

Truss AO:

$$\begin{gathered} \frac{\partial v(x)}{\partial x} = u_o \Phi'_{AO}(x) \sin(pt+\theta) = \\ \lambda_1 \sqrt{p} \left[C_1^{AO} \cos(\lambda_1 \sqrt{p} x) - C_2^{AO} \sin(\lambda_1 \sqrt{p} x) + C_3^{AO} \mathrm{ch}(\lambda_1 \sqrt{p} x) + C_4^{AO} (\mathrm{sh} \lambda_1 \sqrt{p} x) \right] \sin(pt+\theta) \end{gathered} \tag{A19}$$

$$\begin{gathered} \frac{\partial^2 v(x)}{\partial x^2} = u_o \Phi''_{AO}(x) \sin(pt+\theta) = \\ (\lambda_1 \sqrt{p})^2 \left[-C_1^{AO} \sin(\lambda_1 \sqrt{p} x) - C_2^{AO} \cos(\lambda_1 \sqrt{p} x) + C_3^{AO} \mathrm{sh}(\lambda_1 \sqrt{p} x) + C_4^{AO} \mathrm{ch}(\lambda_1 \sqrt{p} x) \right] \sin(pt+\theta); \end{gathered} \tag{A20}$$

$$\begin{gathered} \frac{\partial^3 v(x)}{\partial x^3} = \Phi'''_{AO}(x) \sin(pt+\theta) = \\ (\lambda_1 \sqrt{p})^3 \left[-C_1^{AO} \cos(\lambda_1 \sqrt{p} x) + C_2^{AO} \sin(\lambda_1 \sqrt{p} x) + C_3^{AO} \mathrm{ch}(\lambda_1 \sqrt{p} x) + C_4^{AO} \mathrm{sh}(\lambda_1 \sqrt{p} x) \right] \sin(pt+\theta) \end{gathered} \tag{A21}$$

$$\frac{\partial \varphi(x)}{\partial x} = \psi'_{AO} \sin(pt+\theta) = \delta_1 p \left[D_1^{AO} \cos(\delta_1 p x) - D_2^{AO} \sin(\delta_1 p x) \right] \sin(pt+\theta) \tag{A22}$$

Truss OD:

$$\frac{\partial v(x)}{\partial x} = \lambda_1 \sqrt{p} \left[C_1^{OD} \cos(\lambda_1 \sqrt{p} x) - C_2^{OD} \sin(\lambda_1 \sqrt{p} x) + C_3^{OD} \mathrm{ch}(\lambda_1 \sqrt{p} x) + C_4^{OD} \mathrm{sh}(\lambda_1 \sqrt{p} x) \right] \sin(pt+\theta) \tag{A23}$$

$$\frac{\partial^2 v(x)}{\partial x^2} = (\lambda_1 \sqrt{p})^2 \left[-C_1^{OD} \sin(\lambda_1 \sqrt{p} x) - C_2^{OD} \cos(\lambda_1 \sqrt{p} x) + C_3^{OD} \mathrm{sh}(\lambda_1 \sqrt{p} x) + C_4^{OD} \mathrm{ch}(\lambda_1 \sqrt{p} x) \right] \sin(pt+\theta) \tag{A24}$$

$$\frac{\partial^3 v(x)}{\partial x^3} = (\lambda_1 \sqrt{p})^3 \left[-C_1^{OD} \cos(\lambda_1 \sqrt{p} x) + C_2^{OD} \sin(\lambda_1 \sqrt{p} x) + C_3^{OD} \mathrm{ch}(\lambda_1 \sqrt{p} x) + C_4^{OD} \mathrm{sh}(\lambda_1 \sqrt{p} x) \right] \sin(pt+\theta) \tag{A25}$$

$$\frac{\partial \varphi(x)}{\partial x} = \varphi_o \psi'_{AO} \sin(pt+\theta) = \varphi_o \delta_1 p \left[D_1^{OD} \cos(\delta_1 px) - D_2^{OD} \sin(\delta_1 px) \right] \sin(pt+\theta) \tag{A26}$$

Truss OB:

$$\frac{\partial v(x)}{\partial x} = \lambda_2 \sqrt{p} \left[C_1^{OB} \cos(\lambda_2 \sqrt{p} x) - C_2^{OB} \sin(\lambda_2 \sqrt{p} x) + C_3^{OB} \mathrm{ch}(\lambda_2 \sqrt{p} x) + C_4^{OB} \mathrm{sh}(\lambda_2 \sqrt{p} x) \right] \sin(pt+\theta) \tag{A27}$$

$$\frac{\partial^2 v(x)}{\partial x^2} = (\lambda_2 \sqrt{p})^2 \left[-C_1^{OB} \sin(\lambda_2 \sqrt{p} x) - C_2^{OB} \cos(\lambda_2 \sqrt{p} x) + C_3^{OB} \mathrm{sh}(\lambda_2 \sqrt{p} x) + C_4^{OB} \mathrm{ch}(\lambda_2 \sqrt{p} x) \right] \tag{A28}$$

$$\frac{\partial^3 v(x)}{\partial x^3} = (\lambda_2 \sqrt{p})^3 \left[-C_1^{OB} \cos(\lambda_2 \sqrt{p} x) + C_2^{OB} \sin(\lambda_2 \sqrt{p} x) + C_3^{OB} \mathrm{ch}(\lambda_2 \sqrt{p} x) + C_4^{OB} \mathrm{sh}(\lambda_2 \sqrt{p} x) \right] \sin(pt+\theta) \tag{A29}$$

$$\frac{\partial \varphi(x)}{\partial x} = \varphi_o \delta_2 p \left[D_1^{OB} \cos(\delta_2 px) - D_2^{OB} \sin(\delta_2 px) \right] \sin(pt+\theta) \tag{A30}$$

Truss OC:

$$\frac{\partial v(x)}{\partial x} = \lambda_2 \sqrt{p} \left[C_1^{OC} \cos(\lambda_2 \sqrt{p} x) - C_2^{OC} \sin(\lambda_2 \sqrt{p} x) + C_3^{OC} \mathrm{ch}(\lambda_2 \sqrt{p} x) + C_4^{OC} \mathrm{sh}(\lambda_2 \sqrt{p} x) \right] \sin(pt+\theta) \tag{A31}$$

$$\frac{\partial^2 v(x)}{\partial x^2} = (\lambda_2 \sqrt{p})^2 \left[-C_1^{OC} \sin(\lambda_2 \sqrt{p} x) - C_2^{OC} \cos(\lambda_2 \sqrt{p} x) + C_3^{OC} \mathrm{sh}(\lambda_2 \sqrt{p} x) + C_4^{OC} \mathrm{ch}(\lambda_2 \sqrt{p} x) \right] \sin(pt+\theta) \tag{A32}$$

$$\frac{\partial^3 v(x)}{\partial x^3} = (\lambda_2 \sqrt{p})^3 \left[-C_1^{OC} \cos(\lambda_2 \sqrt{p} x) + C_2^{OC} \sin(\lambda_2 \sqrt{p} x) + C_3^{OC} \mathrm{ch}(\lambda_2 \sqrt{p} x) + C_4^{OC} \mathrm{sh}(\lambda_2 \sqrt{p} x) \right] \sin(pt+\theta) \tag{A33}$$

$$\frac{\partial \varphi(x)}{\partial x} = \varphi_o \delta_2 p \left[D_1^{AO} \cos(\delta_2 px) - D_2^{AO} \sin(\delta_2 px) \right] \sin(pt+\theta) \tag{A34}$$

The following relations will be obtained:

$$C_2^{AO} + C_4^{AO} = 0 \tag{A35}$$

$$C_1^{AO} + C_3^{AO} = 0 \tag{A36}$$

$$D_2^{AO} = 0 \tag{A37}$$

$$C_1^{OD} \sin(\lambda_1 \sqrt{p} L_1) + C_2^{OD} \cos(\lambda_1 \sqrt{p} L_1) + C_3^{OD} \mathrm{sh}(\lambda_1 \sqrt{p} L_1) + C_4^{OD} \mathrm{ch}(\lambda_1 \sqrt{p} L_1) = 0 \tag{A38}$$

$$C_1^{OD} \cos(\lambda_1 \sqrt{p} L_1) - C_2^{OD} \sin(\lambda_1 \sqrt{p} L_1) + C_3^{OD} \mathrm{ch}(\lambda_1 \sqrt{p} L_1) + C_4^{OD} \mathrm{sh}(\lambda_1 \sqrt{p} L_1) = 0 \tag{A39}$$

$$D_1^{OD} \sin(\delta_1 p L_1) + D_2^{OD} \cos(\delta_1 p L_1) = 0 \tag{A40}$$

$$C_1^{OB} \sin(\lambda_2 \sqrt{p} L_3) + C_2^{OB} \cos(\lambda_2 \sqrt{p} L_3) + C_3^{OB} \mathrm{sh}(\lambda_2 \sqrt{p} L_3) + C_4^{OB} \mathrm{ch}(\lambda_2 \sqrt{p} L_3) = 0 \tag{A41}$$

$$-C_1^{OB} \sin(\lambda_2 \sqrt{p} L_3) - C_2^{OB} \cos(\lambda_2 \sqrt{p} L_3) + C_3^{OB} \mathrm{sh}(\lambda_2 \sqrt{p} L_3) + C_4^{OB} \mathrm{ch}(\lambda_2 \sqrt{p} L_3) = 0 \tag{A42}$$

$$D_1^{OB} \cos(\delta_2 p L_3) - D_2^{OB} \sin(\delta_2 p L_3) = 0 \tag{A43}$$

$$C_1^{OC} \sin(\lambda_2 \sqrt{p} L_3) + C_2^{OC} \cos(\lambda_2 \sqrt{p} L_3) + C_3^{OC} \mathrm{sh}(\lambda_2 \sqrt{p} L_3) + C_4^{OC} \mathrm{ch}(\lambda_2 \sqrt{p} L_3) = 0 \tag{A44}$$

$$-C_1^{OC} \sin(\lambda_2 \sqrt{p} L_3) - C_2^{OC} \cos(\lambda_2 \sqrt{p} L_3) + C_3^{OC} \mathrm{sh}(\lambda_2 \sqrt{p} L_3) + C_4^{OC} \mathrm{ch}(\lambda_2 \sqrt{p} L_3) = 0 \tag{A45}$$

$$D_1^{OC} \cos(\delta_2 p L_3) - D_2^{OC} \sin(\delta_2 p L_3) = 0 \tag{A46}$$

which represent a system of 12 equations.

Appendix B

These nine conditions lead to nine relationships:

$$\begin{gathered} C_1^{AO} \sin(\lambda_1 \sqrt{p} L_2) + C_2^{AO} \cos(\lambda_1 \sqrt{p} L_2) + C_3^{AO} \mathrm{sh}(\lambda_1 \sqrt{p} L_2) + C_4^{AO} \mathrm{ch}(\lambda_1 \sqrt{p} L_2) = \\ C_2^{OD} + C_4^{OD} = C_2^{OB} + C_4^{OB} = C_2^{OC} + C_4^{OC} \end{gathered} \tag{A47}$$

$$\begin{gathered} \lambda_1 \sqrt{p} \left[C_1^{AO} \cos(\lambda_1 \sqrt{p} L_2) - C_2^{AO} \sin(\lambda_1 \sqrt{p} L_2) + C_3^{AO} \mathrm{ch}(\lambda_1 \sqrt{p} L_2) + C_4^{AO} \mathrm{sh}(\lambda_1 \sqrt{p} L_2) \right] = \\ \lambda_1 \sqrt{p} \left[C_1^{OD} + C_3^{OD} \right] = D_2^{OB} = -D_2^{OC} \end{gathered} \tag{A48}$$

$$D_2^{AO} = D_2^{OD} = \lambda_2\sqrt{p}\left[C_1^{OB} + C_3^{OB}\right] = \lambda_2\sqrt{p}\left[C_1^{OC} + C_3^{OC}\right] \tag{A49}$$

$$\text{or}: \ \frac{1}{\lambda_2\sqrt{p}} D_2^{AO} = \frac{1}{\lambda_2\sqrt{p}} D_2^{OD} = C_1^{OB} + C_3^{OB} = C_1^{OC} + C_3^{OC} \tag{A50}$$

in which, together, (A35)–(A46) represent a system of 21 equations.

From (A47), we can obtain the equations:

$$C_2^{OB} + C_4^{OB} - C_2^{OD} - C_4^{OD} = 0 \tag{A51}$$

$$C_2^{OC} + C_4^{OC} - C_2^{OD} - C_4^{OD} = 0 \tag{A52}$$

$$C_1^{AO}\sin(\lambda_1\sqrt{p}L_2) + C_2^{AO}\cos(\lambda_1\sqrt{p}L_2) + C_3^{AO}\text{sh}(\lambda_1\sqrt{p}L_2) + C_4^{AO}\text{ch}(\lambda_1\sqrt{p}L_2) - C_2^{OD} + C_4^{OD} = 0. \tag{A53}$$

From (A48), we can obtain the equations:

$$2\!: C_1^{OB} + C_3^{OB} - \frac{1}{\lambda_2\sqrt{p}} D_2^{AO} = 0 \tag{A54}$$

$$C_1^{OC} + C_3^{OC} - \frac{1}{\lambda_2\sqrt{p}} D_2^{AO} = 0 \tag{A55}$$

$$D_2^{AO} - D_2^{OD} = 0 \tag{A56}$$

From (A49), we obtain:

$$D_2^{OB} - \lambda_1\sqrt{p}\left[C_1^{OD} + C_3^{OD}\right] = 0 \tag{A57}$$

$$D_2^{OC} - \lambda_1\sqrt{p}\left[C_1^{OD} + C_3^{OD}\right] = 0; \tag{A58}$$

$$\lambda_1\sqrt{p}\left[C_1^{AO}\cos(\lambda_1\sqrt{p}L_2) - C_2^{AO}\sin(\lambda_1\sqrt{p}L_2) + C_3^{AO}\text{ch}(\lambda_1\sqrt{p}L_2) + C_4^{AO}\text{sh}(\lambda_1\sqrt{p}L_2)\right] - \lambda_1\sqrt{p}\left[C_1^{OD} + C_3^{OD}\right] = 0 \tag{A59}$$

Appendix C

Three more conditions are needed to obtain the constants in the written differential equations.

They are obtained by considering the equilibrium of an infinitesimal element of the mass containing the point O.

The sum of the four shear forces in O must be zero, so:

$$T1 + T2 + T3 - T4 = 0$$

Replacing the expressions of the shear force determined for the four bars in O, we obtain:

$$\begin{aligned}&\lambda_1^3 E_1 I_{z1}\left(-C_1^{AO}\cos(\lambda_1\sqrt{p}\,L_2) + C_2^{AO}\sin(\lambda_1\sqrt{p}\,L_2) + C_3^{AO}\text{ch}(\lambda_1\sqrt{p}\,L_2)\right.\\ &\left.+ C_4^{AO}\text{sh}(\lambda_1\sqrt{p}\,L_2)\right) + \lambda_1^3 E_1 I_{z1}\left(-C_1^{OD} + C_3^{OD}\right) + \lambda_2^3 E_2 I_{z2}\left(-C_1^{OB} + C_3^{OB}\right)\\ &- \lambda_2^3 E_2 I_{z2}\left(-C_1^{OC} + C_3^{OC}\right) = 0\end{aligned} \tag{A60}$$

With the notation:

$$a_1 = \frac{E_2 I_{p2}}{E_1 I_{z1}}\left(\frac{\lambda_2}{\lambda_1}\right)^3 \tag{A61}$$

we can write:

$$\begin{aligned}&\left(-C_1^{AO}\cos(\lambda_1\sqrt{p}\,L_2) + C_2^{AO}\sin(\lambda_1\sqrt{p}\,L_2) + C_3^{AO}\text{ch}(\lambda_1\sqrt{p}\,L_2)\right.\\ &\left.+ C_4^{AO}\text{sh}(\lambda_1\sqrt{p}\,L_2)\right) + \left(-C_1^{OD} + C_3^{OD}\right) + a_1\left(-C_1^{OB} + C_3^{OB}\right) - a_1\left(-C_1^{OC} + C_3^{OC}\right) = 0\end{aligned} \tag{A62}$$

A similar equilibrium relation is obtained for bending and torsional moments.

$$M^{b1} + M^{b2} + M^{t3} - M^{t4} = 0$$

$$M^{t1} + M^{t2} + M^{b3} - M^{b4} = 0$$

Taking into account relations (27)–(43), we obtain:

$$-E_1 I_{z1} \frac{\partial^2 \Phi_{AO}(L_2)}{\partial x^2} - E_1 I_{z1} \frac{\partial^2 \Phi_{OD}(0)}{\partial x^2} + G_2 I_{p2} \frac{\partial \varphi_{OB}(0)}{\partial x} - G_2 I_{p2} \frac{\partial \varphi_{OC}(0)}{\partial x} = 0 \quad \text{(A63)}$$

$$G_1 I_{p1} \frac{\partial \varphi_{AO}(L_2)}{\partial x} + G_1 I_{p1} \frac{\partial \varphi_{OD}(0)}{\partial x} - E_2 I_{z2} \frac{\partial^2 \Phi_{OB}(L_2)}{\partial x^2} + E_2 I_{z2} \frac{\partial^2 \Phi_{OC}(0)}{\partial x^2} = 0 \quad \text{(A64)}$$

Using (11)–(14) and (23)–(26), the results are:

$$\begin{gathered} C_1^{AO} \sin(\lambda_1 \sqrt{p}\, L_2) + C_2^{AO} \cos(\lambda_1 \sqrt{p}\, L_2) - C_3^{AO} \mathrm{sh}(\lambda_1 \sqrt{p}\, L_2) - C_4^{AO} \mathrm{ch}(\lambda_1 \sqrt{p}) - \\ (-C_2^{OD} + C_4^{OD}) + \frac{G_2 I_{p2}}{E_1 I_{z1}} \frac{\delta_2}{\lambda_1^2} (D_1^{OB} - D_1^{OC}) = 0 \end{gathered} \quad \text{(A65)}$$

$$\begin{gathered} \frac{G_1 I_{p1}}{E_2 I_{z2}} \frac{\delta_1}{\lambda_2^2} (D_1^{AO} \cos(\delta_1 p L_2) - D_2^{AO} \sin(\delta_1 p L_2)) + \frac{G_1 I_{p1}}{E_2 I_{z2}} \frac{\delta_1}{\lambda_2^2} D_1^{OD} - \\ (-C_2^{OB} + C_4^{OB}) + (-C_2^{OC} + C_4^{OC}) = 0 \end{gathered} \quad \text{(A66)}$$

This denoted:

$$a_2 = \frac{G_2 I_{p2}}{E_1 I_{z1}} \frac{\delta_2}{\lambda_1^2},\ a_3 = \frac{G_1 I_{p1}}{E_2 I_{z2}} \frac{\delta_1}{\lambda_2^2} \quad \text{(A67)}$$

$$\begin{gathered} C_1^{AO} \sin(\lambda_1 \sqrt{p}\, L_2) + C_2^{AO} \cos(\lambda_1 \sqrt{p}\, L_2) - C_3^{AO} \mathrm{sh}(\lambda_1 \sqrt{p}\, L_2) - C_4^{AO} \mathrm{ch}(\lambda_1 \sqrt{p}\, L_2) - \\ \left(-C_2^{OD} + C_4^{OD}\right) + a_2 \left(D_1^{OB} - D_1^{OC}\right) = 0 \end{gathered} \quad \text{(A68)}$$

$$a_3 \left(D_1^{AO} \cos(\delta_1 p L_2) - D_2^{AO} \sin(\delta_1 p L_2)\right) + a_3 D_1^{OD} - \left(-C_2^{OB} + C_4^{OB}\right) + \left(-C_2^{OC} + C_4^{OC}\right) = 0 \quad \text{(A69)}$$

This represents a system with 24 equations with 24 unknowns. From the system formed, we must determine the constants C_1^{AO}, C_2^{AO}, C_3^{AO}, C_4^{AO}, C_1^{OD}, C_2^{OD}, C_3^{OD}, C_4^{OD}, C_1^{OB}, C_2^{OB}, C_3^{OB}, C_4^{OB}, C_1^{OC}, C_2^{OC}, C_3^{OC}, C_4^{OC}, D_1^{AO}, D_2^{AO}, D_1^{OD}, D_2^{OD}, D_1^{OB}, D_2^{OB}, D_1^{OC}, D_2^{OC}. They will be denoted, in the following:

$$\{B\} = \left\{ \begin{array}{c} \{B^{AO}\} \\ \{B^{OD}\} \\ \{B^{OB}\} \\ \{B^{OC}\} \end{array} \right\} \quad \text{(A70)}$$

with:

$$\left\{B^{AO}\right\} = [\ C_1^{AO} \quad C_2^{AO} \quad C_3^{AO} \quad C_4^{AO} \quad D_1^{AO} \quad D_2^{AO}\ ] \quad \text{(A71)}$$

$$\left\{B^{OD}\right\} = [\ C_1^{OD} \quad C_2^{OD} \quad C_3^{OD} \quad C_4^{OD} \quad D_1^{OD} \quad D_2^{OD}\ ] \quad \text{(A72)}$$

$$\left\{B^{OB}\right\} = [\ C_1^{OB} \quad C_2^{OB} \quad C_3^{OB} \quad C_4^{OB} \quad D_1^{OB} \quad D_2^{OB}\ ] \quad \text{(A73)}$$

$$\left\{B^{OC}\right\} = [\ C_1^{OC} \quad C_2^{OC} \quad C_3^{OC} \quad C_4^{OC} \quad D_1^{OC} \quad D_2^{OC}\ ] \quad \text{(A74)}$$

A homogeneous linear system was obtained. In order to have other solutions besides the trivial solution zero, the determinant of the system must be zero. Putting this condition, the obtained eigenfrequencies of the system can be determined from the obtained equation.

Denote:

$$[A_{11}] = \begin{bmatrix} 0 & 1 & 0 & 1 & 0 & 0 \\ 1 & 0 & 1 & 0 & 0 & 0 \\ 0 & 0 & 0 & 0 & 0 & 1 \\ \sin(\lambda_2\sqrt{p}L_3) & \cos(\lambda_2\sqrt{p}L_3) & sh(\lambda_2\sqrt{p}L_3) & ch(\lambda_2\sqrt{p}L_3) & 0 & 0 \\ -\sin(\lambda_2\sqrt{p}L_3) & -\cos(\lambda_2\sqrt{p}L_3) & \mathrm{sh}(\lambda_2\sqrt{p}L_3) & \mathrm{ch}(\lambda_2\sqrt{p}L_3) & 0 & 0 \\ 0 & 0 & 0 & 0 & \cos(\delta_2 pL_3) & -\sin(\delta_2 pL_3) \end{bmatrix} \quad \text{(A75)}$$

$$[A_{12}] = \begin{bmatrix} 0 & 0 & 0 & 0 & 0 & 0 & 0 & -1 & 0 & -1 & 0 & 0 \\ 0 & 0 & 0 & 0 & 0 & -\frac{1}{\lambda_1\sqrt{p}} & 0 & 0 & 0 & 0 & 0 & 0 \\ 0 & 0 & 0 & 0 & 0 & 0 & -\lambda_1\sqrt{p} & 0 & -\lambda_1\sqrt{p} & 0 & 0 & 0 \\ 0 & 0 & 0 & 0 & 0 & 0 & 0 & 0 & 0 & 0 & 0 & 0 \\ 0 & 0 & 0 & 0 & 0 & 0 & 0 & 0 & 0 & 0 & 0 & 0 \\ 0 & 0 & 0 & 0 & 0 & 0 & 0 & 0 & 0 & 0 & 0 & 0 \end{bmatrix} \quad \text{(A76)}$$

$$[A_{21}] = \begin{bmatrix} 0 & 0 & 0 & 0 & 0 & 0 & 0 & 0 & 0 & 0 & 0 & 0 \\ 0 & 0 & 0 & 0 & 0 & 0 & 0 & 0 & 0 & 0 & 0 & 0 \\ 0 & 0 & 0 & 0 & 0 & 0 & 0 & 0 & 0 & 0 & 0 & 0 \\ 0 & 0 & 0 & 0 & 0 & 0 & 0 & 0 & 0 & 0 & 0 & 0 \\ 0 & 0 & 0 & 0 & 0 & 0 & 0 & 0 & 0 & 0 & 0 & 0 \\ 0 & 0 & 0 & 0 & 0 & 0 & 0 & 0 & 0 & 0 & 0 & 0 \\ 0 & 0 & 0 & 0 & 0 & 0 & 0 & 0 & 0 & 0 & 0 & 0 \\ 0 & 0 & 0 & 0 & 0 & 0 & 0 & 0 & 0 & 0 & 0 & 0 \\ 0 & 0 & 0 & 0 & 0 & 0 & 0 & 0 & 0 & 0 & 0 & 0 \\ -a_1 & 0 & -a_1 & 0 & 0 & 0 & a_1 & 0 & -a_1 & 0 & 0 & 0 \\ 0 & 0 & 0 & 0 & a_2 & 0 & 0 & 0 & 0 & 0 & -a_2 & 0 \\ 0 & 1 & 0 & -1 & 0 & 0 & 0 & -1 & 0 & 1 & 0 & 0 \end{bmatrix} \quad \text{(A77)}$$

In the following, we denote:

$$[A_{22}] = \begin{bmatrix} A_{22}a & A_{22}b \\ A_{22}c & A_{22}d \end{bmatrix} \quad \text{(A78)}$$

where:

$$[A_{22}a] = \begin{bmatrix} 0 & 1 & 0 & 1 & 0 & 0 \\ 1 & 0 & 1 & 0 & 0 & 0 \\ 0 & 0 & 0 & 0 & 0 & 1 \\ 0 & 0 & 0 & 0 & 0 & 0 \\ 0 & 0 & 0 & 0 & 0 & 0 \\ 0 & 0 & 0 & 0 & 0 & 0 \end{bmatrix} \quad \text{(A79)}$$

$$[A_{12}b] = \begin{bmatrix} 0 & 0 & 0 & 0 & 0 & 0 \\ 0 & 0 & 0 & 0 & 0 & 0 \\ 0 & 0 & 0 & 0 & 0 & 0 \\ \sin(\lambda_1\sqrt{p}L_1) & \cos(\lambda_1\sqrt{p}L_1) & \mathrm{sh}(\lambda_1\sqrt{p}L_1) & \mathrm{ch}(\lambda_1\sqrt{p}L_1) & 0 & 0 \\ \cos(\lambda_1\sqrt{p}L_1) & -\sin(\lambda_1\sqrt{p}L_1) & \mathrm{ch}(\lambda_1\sqrt{p}L_1) & \mathrm{sh}(\lambda_1\sqrt{p}L_1) & 0 & 0 \\ 0 & 0 & 0 & 0 & \sin(\delta_1 pL_1) & \cos(\delta_1 pL_1) \end{bmatrix} \quad \text{(A80)}$$

$$[A_{12}d] = \begin{bmatrix} 0 & -1 & 0 & 1 & 0 & 0 \\ 0 & 0 & 0 & 0 & 0 & -1 \\ -1 & 0 & -1 & 0 & 0 & 0 \\ -1 & 0 & 1 & 0 & 0 & 0 \\ 0 & 1 & 0 & -1 & 0 & 0 \\ 0 & 0 & 0 & 0 & \mathrm{a}_3 & 0 \end{bmatrix} \quad \text{(A81)}$$

$$[A_{12}c] = \begin{bmatrix} \sin(\lambda_1\sqrt{p}L_2) & \cos(\lambda_1\sqrt{p}L_2) & \mathrm{sh}(\lambda_1\sqrt{p}L_2) & \mathrm{ch}(\lambda_1\sqrt{p}L_2) & 0 & 0 \\ 0 & 0 & 0 & 0 & 0 & 1 \\ \cos(\lambda_1\sqrt{p}L_2) & -\sin(\lambda_1\sqrt{p}L_2) & \mathrm{ch}(\lambda_1\sqrt{p}L_2) & \mathrm{sh}(\lambda_1\sqrt{p}L_2) & 0 & 0 \\ -\cos(\lambda_1\sqrt{p}L_2 & \sin(\lambda_1\sqrt{p}L_2) & \mathrm{ch}(\lambda_1\sqrt{p}L_2) & \mathrm{sh}(\lambda_1\sqrt{p}L_2) & 0 & 0 \\ \sin(\lambda_1\sqrt{p}L_2 & \cos(\lambda_1\sqrt{p}L_2) & -\mathrm{sh}(\lambda_1\sqrt{p}L_2) & -\mathrm{ch}(\lambda_1\sqrt{p}L_2) & 0 & 0 \\ 0 & 0 & 0 & 0 & \mathrm{a}_3\cos(\delta_1 pL_2) & -\mathrm{a}_3\sin(\delta_1 pL_2) \end{bmatrix} \quad \text{(A82)}$$

In this case, the matrix of the system can be written, in a concise form:

$$A_{24\times 24} = \begin{bmatrix} A_{11} & 0 & A_{12} \\ 0 & A_{11} & A_{12} \\ A_{21} & A_{21} & A_{22} \end{bmatrix} \quad \text{(A83)}$$

and the system becomes:

$$[A]\{B\} = \{0\} \quad \text{(A84)}$$

The condition:

$$\det(A) = 0 \quad \text{(A85)}$$

offers the eigenvalues of the system of differential Equations (6)–(9) and (19)–(22).

Appendix D

Let us now consider one of the identical trusses OB and OC. The truss is clamped in O and supported in B (or C). Equation (1) is also valid for the OB bar (OC), with the boundary conditions:

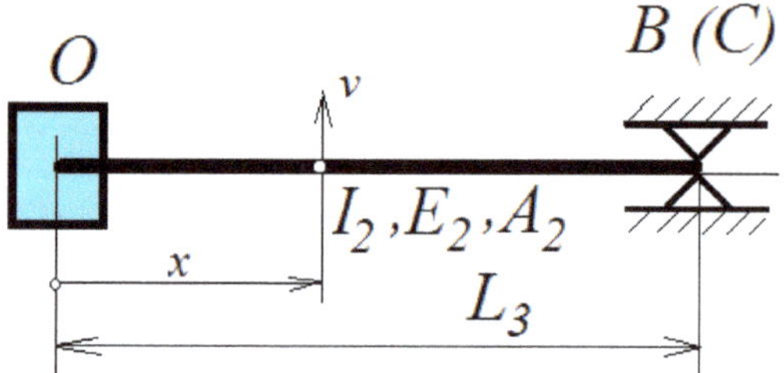

Figure A1. Model of a single truss.

For point O:

$$x = 0;\ v(0,t) = 0;\ v'(0,t) = 0;\ \varphi(0,t) = 0 \quad \text{(A86)}$$

and for the point B(C):

$$x = L_3;\ v(L_3,t) = 0;\ M^b(L_3,t) = 0;\ M^t(L_3,t) = 0 \quad \text{(A87)}$$

The solution:

$$v(x,t) = \Phi(x)\sin(pt+\theta) \quad \text{(A88)}$$

offers the new equation:

$$\frac{\partial^4\Phi}{\partial x^4} - p^2\frac{\rho_2 A_2}{E_2 I_{z2}}\Phi = 0 \quad \text{(A89)}$$

If noted:

$$\lambda_2^4 = \frac{\rho_2 A_2}{E_2 I_{z2}} \quad \text{(A90)}$$

the solution is:

$$\Phi(x) = C_1\sin(\lambda_2\sqrt{p}x) + C_2\cos(\lambda_2\sqrt{p}x) + C_3\mathrm{sh}(\lambda_2\sqrt{p}x) + C_4\mathrm{ch}(\lambda_2\sqrt{p}x) \quad \text{(A91)}$$

Torsional vibrations are described by Equation (15). By choosing φ as:

$$\varphi(x,t) = \psi(x)\sin(pt+\theta) \quad \text{(A92)}$$

and introducing in Equation (15), we obtain:

$$\frac{\partial^2\psi}{\partial x^2} + p^2\delta_2^2\psi = 0 \quad \text{(A93)}$$

The solution of the differential Equation (A93) will be:

$$\psi(x) = D_1\sin(\delta_2 px) + D_2\cos(\delta_2 px) \quad \text{(A94)}$$

References

1. Meirovitch, L. *Analytical Methods in Vibrations*; McMillan: Hong Kong, China, 1967.
2. Mihalcica, M.; Paun, M.; Vlase, S. The Use of Structural Symmetries of a U12 Engine in the Vibration Analysis of a Transmission. *Symmetry* **2019**, *11*, 1296. [CrossRef]
3. Shi, C.Z.; Parker, R.G. Modal structure of centrifugal pendulum vibration absorber systems with multiple cyclically symmetric groups of absorbers. *J. Sound Vib.* **2013**, *332*, 4339–4353. [CrossRef]
4. Paliwal, D.N.; Pandey, R.K. Free vibrations of circular cylindrical shell on Winkler and Pasternak foundations. *Int. J. Press. Vessel. Pip.* **1996**, *69*, 79–89. [CrossRef]
5. Celep, Z. On the axially symmetric vibration of thick circular plates. *Ing. Archiv.* **1978**, *47*, 411–420. [CrossRef]
6. Zingoni, A. A group-theoretic finite-difference formulation for plate eigenvalue problems. *Comput. Struct.* **2012**, *112–113*, 266–282. [CrossRef]
7. Vlase, S.; Marin, M.; Oechsner, A. Considerations of the transverse vibration of a mechanical system with two identical bars. *Proc. Inst. Mech. Eng. Part L J. Mater. Des. Appl.* **2019**, *233*, 1318–1323. [CrossRef]
8. Holm, D.D.; Stoica, C.; Ellis, D.C.P. *Geometric Mechanics and Symmetry*; Oxford University Press: Oxford, UK, 2009.
9. Marsden, J.E.; Ratiu, T.S. *Introduction to Mechanics and Symmetry: A Basic Exposition of Classical Mechanical Systems*; Springer: Berlin, Germany, 2003; p. 586.
10. Singer, S.F. *Symmetry in Mechanics*; Springer: Berlin, Germany, 2004.
11. Zavadskas, E.K.; Bausys, R.; Antucheviciene, J. Civil Engineering and Symmetry. *Symmetry* **2019**, *11*, 501. [CrossRef]
12. Ganghoffer, J.-F.; Mladenov, I. *Similarity, Symmetry and Group Theoretical Methods in Mechanics*; Lectures at the International Centre for Mechanical Sciences; Springer: Berlin, Germany, 2015.
13. Lin, J.; Jin, S.; Zheng, C.; Li, Z.M.; Liu, Y.H. Compliant assembly variation analysis of aeronautical panels using unified substructures with consideration of identical parts. *Comput. Aided Des.* **2014**, *57*, 29–40. [CrossRef]
14. Elkin, V.I. General-Solutions of Partial-Differential Equation Systems with Identical Principal Parts. *Differ. Equ.* **1985**, *21*, 952–959.
15. Menshikh, O.F. Traveling Waves of a System of Quasilinear Equations with Identical Principal Parts. *Dokl. Akad. Nauk USSR* **1976**, *227*, 555–557.
16. Zingoni, A. On the best choice of symmetry group for group-theoretic computational schemes in solid and structural mechanics. *Comput. Struct.* **2019**, *223*, 106101. [CrossRef]
17. Zingoni, A. Group-theoretic insights on the vibration of symmetric structures in engineering. *Philos. Trans. R. Soc. A Math. Phys. Eng. Sci.* **2014**, *372*, 20120037. [CrossRef]
18. Amaral, R.R.; Troina, G.S.; Fragassa, C.; Pavlovic, A.; Cunha, M.L.; Rocha, L.A.; dos Santos, E.D.; Isoldi, L.A. Constructal design method dealing with stiffened plates and symmetry boundaries. *Theor. Appl. Mech.* **2020**, *10*, 366–376. [CrossRef]
19. Scutaru, M.L.; Vlase, S.; Marin, M.; Modrea, A. New analytical method based on dynamic response of planar mechanical elastic systems. *Bound. Value Probl.* **2020**, *2020*, 104. [CrossRef]
20. Chen, Y.; Feng, J. Group-Theoretic Exploitations of Symmetry in Novel Prestressed Structures. *Symmetry* **2018**, *10*, 229. [CrossRef]
21. Harth, P.; Beda, P.; Michelberger, P. Static analysis and reanalysis of quasi-symmetric structure with symmetry components of the symmetry groups C-3v and C-1v. *Eng. Struct.* **2017**, *152*, 397–412. [CrossRef]
22. He, J.H.; Latifizadeh, H. A general numerical algorithm for nonlinear differential equations by the variational iteration method. *Int. J. Numer. Methods Heat Fluid Flow* **2020**, *30*, 4797–4810. [CrossRef]
23. He, C.H.; Liu, C.; Gepreel, K.A. Low frequency property of a fractal vibration model for a concrete beam. *Fractals* **2021**. [CrossRef]
24. Ali, M.; Anjum, N.; Ain, Q.T.; He, J.H. Homotopy Perturbation Method for the Attachment Oscillator Arising in Nanotechnology. *Fibers Polym.* **2021**, *22*. [CrossRef]
25. He, J.-H.; Hou, W.-F.; Qie, N.; Gepreel, K.A.; Shirazi, A.H.; Sedighi, H.M. Hamiltonian-based frequency-amplitude formulation for nonlinear oscillators. *Facta Univ. Ser. Mech. Eng.* **2021**. [CrossRef]
26. Den Hartog, J.P. *Mechanical Vibrations*; Dover Publications: Mineola, NY, USA, 1985.
27. Thorby, D. *Structural Dynamics and Vibrations in Practice: An Engineering Handbook*; CRC Press: Boca Raton, FL, USA, 2012.

28. Timoshenko, P.S.; Gere, J.M. *Theory of Elastic Stability*, 2nd ed.; McGraw-Hill: New York, NY, USA; London, UK, 2009.
29. Tyn, M.U. *Ordinary Differential Equations*; Elsevier: Amsterdam, The Netherlands, 1977.
30. Vlase, S.; Marin, M.; Scutaru, M.L.; Munteanu, R. Coupled transverse and torsional vibrations in a mechanical system with two identical beams. *AIP Adv.* **2017**, *7*, 065301. [CrossRef]
31. Vlase, S.; Marin, M.; Öchsner, A.; Scutaru, M.L. Motion equation for a flexible one-dimensional element used in the dynamical analysis of a multibody system. *Contin. Mech. Thermodyn.* **2019**, *31*, 715–724. [CrossRef]
32. Bhatti, M.M.; Marin, M.; Zeeshan, A.; Ellahi, R.; Abdelsalam, S.I. Swimming of Motile Gyrotactic Microorganisms and Nanoparticles in Blood Flow through Anisotropically Tapered Arteries. *Front. Phys.* **2020**, *8*, 95. [CrossRef]
33. Nicolescu, A.E.; Bobe, A. Weak Solution of Longitudinal Waves in Carbon Nanotubes. *Contin. Mech. Therm.* **2021**, 1–9. [CrossRef]
34. Vlase, S.; Paun, M. Vibration analysis of a mechanical system consisting of two identical parts. *Rom. J. Tech. Sci. Appl. Mech.* **2015**, *60*, 216–230.

Article

Evaluating the Impact of Different Symmetrical Models of Ambient Assisted Living Systems

Wael Alosaimi [1], Md Tarique Jamal Ansari [2], Abdullah Alharbi [1], Hashem Alyami [3], Adil Hussain Seh [4], Abhishek Kumar Pandey [4], Alka Agrawal [4] and Raees Ahmad Khan [4,*]

1 Department of Information Technology, College of Computers and Information Technology, Taif University, P.O. Box 11099, Taif 21944, Saudi Arabia; w.osaimi@tu.edu.sa (W.A.); amharbi@tu.edu.sa (A.A.)
2 Department of Computer Application, Integral University, Lucknow 226026, India; tjtjansari@gmail.com
3 Department of Computer Science, College of Computers and Information Technology, Taif University, P.O. Box 11099, Taif 21944, Saudi Arabia; hyami@tu.edu.sa
4 Department of Information Technology, Babasaheb Bhimrao Ambedkar University, Lucknow 226025, India; ahseh.rs@bbau.ac.in (A.H.S.); abhishekkumarpanday5@gmail.com (A.K.P.); dralka@bbau.ac.in (A.A.)
* Correspondence: khanraees@bbau.ac.in or khanraees@yahoo.com

Abstract: In recent years, numerous attempts have been made to enhance the living standard for old-aged people. Ambient Assisted Living (AAL) is an evolving interdisciplinary field aimed at the exploitation of knowledge and communication technology in health and tele-monitoring systems to combat the impact of the growing aging population. AAL systems are designed for customized, responsive, and predictive requirements, requiring high performance of functionality to ensure interoperability, accessibility, security, and consistency. Standardization, continuity, and assistance of system development have become an urgent necessity to meet the increasing needs for sustainable systems. In this article, we examine and address the methods of the different AAL systems. In addition, we analyzed the acceptance criteria of the AAL framework intending to define and evaluate different AAL-based symmetrical models, leveraging performance characteristics under the integrated fuzzy environment. The main goal is to provide an understanding of the current situation of the AAL-oriented setups. Our vision is to investigate and evaluate the potential symmetrical models of AAL systems and frameworks for the implementation of effective new installations.

Keywords: ambient assisted living; AAL; ambient intelligence; assisted living; user-interfaces; fuzzy logic

Citation: Alosaimi, W.; Ansari, M.T.J.; Alharbi, A.; Alyami, H.; Seh, A.H.; Pandey, A.K.; Agrawal, A.; Khan, R.A. Evaluating the Impact of Different Symmetrical Models of Ambient Assisted Living Systems. *Symmetry* **2021**, *13*, 450. https://doi.org/10.3390/sym13030450

Academic Editor: Jan Awrejcewicz

Received: 10 February 2021
Accepted: 6 March 2021
Published: 10 March 2021

1. Introduction

The current digital environment, comprising smart home products, mobile devices, smart watches, and software applications, has had a significant impact on human lifestyles. These systems have provided a great deal of power to individuals, thereby significantly lowering dependence on others. These advanced devices have not only transformed lifestyles, but have also revolutionized almost every area of human existence. The idea of Ambient Assisted Living (AAL) resulted from these smart technologies, and represents the response to the task of maintaining the standard of living of elderly people. Ambient Assisted Living (AAL) offers a system consisting of smart phones, medical sensors, cellular networks, computers, and health tracking apps [1]. AAL can also be used for different reasons, such as preventing, treating, and enhancing the well-being and health of elderly people.

AAL strives to promote the protection and wellbeing of elderly people and to increase the number of years that elderly people can live comfortably in an area of their own convenience [2–5]. It also reduces the amount of anticipated costs by empowering patients to monitor their serious medical symptoms. AAL is a sub-part of Ambient Intelligence that includes the utilization of ambient intelligent strategies, processes, and technology to allow aged people to survive comfortably for as long as humanly possible despite behavioral problems.

In addition, modern developments in mobile and portable sensors have contributed to realizing the vision of AAL [6,7]. Recent popular electronic applications are fitted with smart configurations, such as accelerometers, navigation systems, GPS, and many other systems that can be used to monitor user mobility. Furthermore, recent developments in digital and sensor technology are promising a new age in health sensors [8]. Scientists and researchers have created discreet sensors in the form of covers, small holter-type gadgets, mobile systems, and smart clothing for tracking health indicators. For instance, blood sugar, blood pressure and heart performance can be evaluated by means of smart technologies such as infrared or photographic sensors. Many measures, such as electroencephalography (EEG), also involve invasive devices, including electrodes. The following Figure 1 shows the graphical representation of an Ambient Assisted Living (AAL) system.

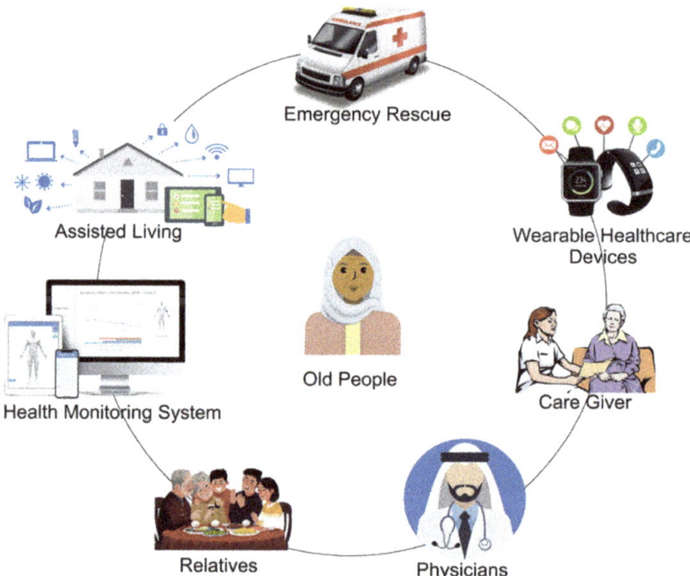

Figure 1. Graphical representation of an Ambient Assisted Living (AAL) system.

This study evaluates different Ambient Assisted Living system's symmetrical models based on the taxonomy adopted from Amina et al. [9]. The study uses integrated fuzzy Analytic Hierarchy Process-Technique for Order Performance by Similarity to Ideal Solution (AHP-TOPSIS), which is a popular multiple-criteria decision-making (MCDM) approach. To handle the complexity of evaluating the efficiency of various Ambient Assisted Living System Symmetrical Models on one parameter, or on the value of another high-precision parameter, the Analytic Hierarchy Process (AHP) method has been extensively used by numerous authors and practitioners. Ghodsypour and O'Brien [10] claim that AHP is much more reliable than other symmetrical models of scoring for analysis procedures. Conceivably, the technique is suitable because the decision-making process has a one-way hierarchical relationship between decision-making stages. Interestingly, Carney and Wallnau [11] noted that the selection parameters for alternatives are not necessarily independent of each other, but rather interconnect. In such a complicated setting, an incorrect result may be obtained. TOPSIS (Technique for Order Performance by Similarity to Idea Solution) [12] is also an appropriate approach for solving MCDM problems. TOPSIS is initiated on the principle that the optimum alternative should also have the smallest distance from the positive idea solution (PIS), and the greatest distance from the negative idea solution. The principle of TOPSIS is logical and comprehensible, and the related calculation is straightforward. Consequently, it is important to note the inherent complexity of specifying accurate subjective opinions to the parameters.

In the subsequent sections of this article, we classify the discussion into five sections. Section 2 presents the different related works. Section 3 discusses the materials and methods used in this paper. Section 4 presents the statistical results and evaluates their quality characteristics according to different metrics. Finally, Section 5 concludes the paper.

2. Related Works

Fuzzy multiple-criteria decision-making (MCDM) provides successful outcomes in the resolution of selection-based problems [13]. The approach has been chosen by several researchers because it can effectively handle the knowledge that is analyzed using a multi-resourced linguistic and quantitative decision-making challenge, and evidence that is diverse.

Lin et al. [14] suggested a fuzzy-based strategy for the selection of an effective smart technology framework for fall detection. To address this problem, a fuzzy collective intelligence methodology was used. Alpha-cut procedures were used in a fuzzy collective intelligence method to determine the fuzzy weights of the parameters with each decision-maker. Then, the fuzzy combination was used to sum the fuzzy weights generated by each decision-maker. Consequently, a fuzzy order preference strategy comparable to the ideal solution was used to assess the appropriateness of a smart technology system for fall detection.

Samanlioglu et al. [15] proposed an approach to choosing the best employee applicant for an IT organization by combining the fuzzy analytic hierarchical process method (fuzzy AHP), including Chang's scale analysis, in addition to fuzzy TOPSIS. The decision-makers' (DMs) verbal assessments were included in the analysis using intuitive fuzzy numbers. They first calculated the value of thirty sub-criteria weights using fuzzy AHP and then, using fuzzy TOPSIS, five IT employees' alternatives were evaluated based on fuzzy AHP weights.

Anand and Vinodh [16] presented research to evaluate Additive Manufacturing (AM) procedures for micro-manufacturing using combined fuzzy AHP-TOPSIS. Parameters weights were derived using fuzzy AHP, whereas rankings were obtained using fuzzy TOPSIS.

Nazam et al. [17] suggested hybrid fuzzy AHP analysis to measure the weight of threat parameters and sub-criteria, in addition to the order efficiency technique using the ideal solution (TOPSIS) procedure to rank and evaluate the risks resulting from the adoption of green supply chain management (GSCM) practices in a fuzzy setting. Their proposed fuzzy risk-oriented assessment theory was implemented for the realistic case of the textile automotive industry. Ultimately, the conceptual model allows academics and professionals to consider the value of performing effective risk assessments while introducing green supply chain interventions.

Ansari et al. [18] addressed the viewpoint of security professionals on the relative value of parameters for the selection of an appropriate Security Requirements Engineering (SRE) system using the multi-criteria decision-making approach. The research was designed and conducted to determine the most suitable SRE approach for quality and stable application development based on the expertise and experience of the security specialist. The hierarchical analysis was performed using a fuzzy TOPSIS method. The efficient SRE selection process was evaluated in pairs.

Kumar et al. [19] used an approach that involves the combination of fuzzy AHP with fuzzy TOPSIS processes to determine the effect of various malware detection techniques in the context of web applications. This research used different variants of a university's software system to test the effect of a variety of current malware detection techniques.

Alenezi et al. [20] used the combined approach of fuzzy AHP-TOPSIS to test security architecture strategies and their attributes. The efficacy of this method was also evaluated on the real-time software system of BBAU, Lucknow, India. In addition, various university web apps were also used to support the findings produced.

The uncertainty associated with Alzheimer's disease complicates the development of patient monitoring software. To address this problem, Coronato described a systematic approach to modelling and developing intelligent Ambient Assisted Living Systems to monitor the behavior of people with cognitive impairments. The major aspect was an automated method to detect irrational behavior, in addition to a technique for the design of secure AAL systems specifically aimed at individuals with cognitive or mental illnesses. The author also implemented their approach for modelling a monitoring system that can demonstrate feasibility through the identification of abnormal circumstances [21].

Koleva et al. highlighted the main obstacles in formulating and executing an efficient AAL system based on their expertise in the eWALL project. They also provided suggestions for solutions to these obstacles [22].

Many surveys and reviews have been conducted that address the functionality of different models of Ambient Assisted Living systems. However, the fuzzy AHP-TOPSIS approach has not been used to evaluate the impact of different symmetrical models of Ambient Assisted Living systems.

3. Materials and Methods

Recent developments in a variety of technical fields have helped to realize the potential of AAL. These innovations involve smart homes, help robots, e-textiles, and portable and implantable sensors. In the following sections, the proposed research methodology is discussed in more detail. Figure 2 shows a functional diagram of the research methodology.

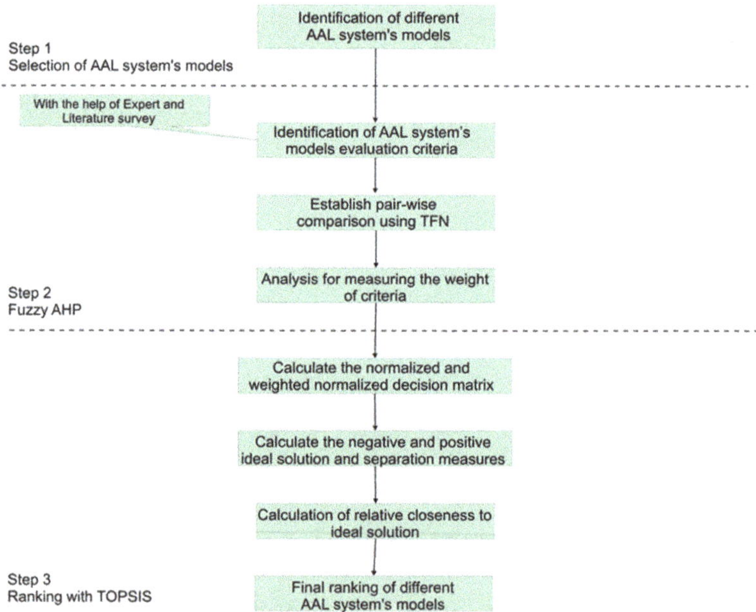

Figure 2. Functional diagram of the research methodology.

3.1. Step 1: Identification of Different AAL System's Models

A variety of ventures have adopted the AAL scope as a field of research. Various references, symmetrical models, systems, and interfaces have been proposed for an acceptable AAL scheme, but few of these have been generally accepted. In this section, we analyze the most comprehensive Ambient Assisted Living system's symmetrical models. A graphical representation of the hierarchy can be seen in Figure 3.

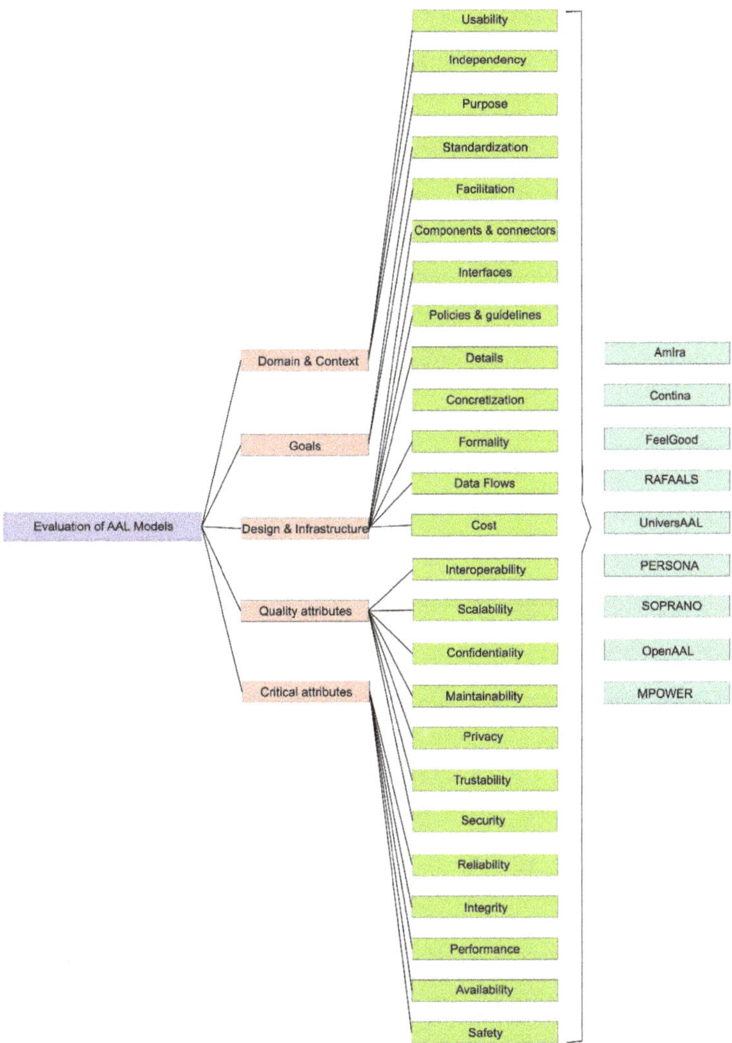

Figure 3. Structure for evaluation of different Ambient Assisted Living system's symmetrical models.

A. *Ambient Intelligence Reference Architecture (AmIRA)*

AmIRA includes an emphasis on processes, structures, and components [23]. It assimilates multi-layer application architecture. According to the given specifications of ambient intelligent systems, the intention of AmIRA is to encourage the re-use of Ambient Intelligence (AmI) processes throughout various AmI systems [24]. The advantage of this design is that each layer is autonomous; however, there is a demand for resources from other layers. AmIRA is uncertain because there are no restrictions mentioned, however, the framework components do not adequately define the business processes.

B. *Continua*

Continua is an attempt to provide compatibility in the area of personal telemedicine. The design identifies different steps and reference system classes [25]. Continua made an explicit option to build a framework that incorporates several domains. It is known to be the only approach that provides a uniform security system. Continua reflects the reference architecture mostly as a symmetrical layered model. Each link is composed to its counterparts by a platform that has been identified. Reference system classes and platforms are the maximum representation level in the Continua design. Conversely, it depends on the execution of the design and lacks any applications.

C. *FeelGood*

FeelGood is an initiative that introduces new architectural elements to complement existing practices by specifying the ecosystem level of service components [26]. The primary objective was to enhance the quality of life in Finland. The suggested RA does not describe comprehensive interfaces, but offers interface specifications and refers to the applicable requirements that could be implemented. The purpose of this RA was to direct the development of the product. Stakeholders and comprehensive services were identified.

D. *RAFAALS*

RAFAALS relates to the reference design for AAL systems [27,28] as a SOA-based symmetrical model of AAL. The key architecture is based on the principle of separating the functionality and the configuration of the data flow among the layers of the process. It facilitates the transfer of events between flexible software applications wherein its modules allow use of little to no awareness of other elements. This structure distinguishes the activities within each evolved producer. The layers are analytically isolated from each other and clearly specified. This is a traditional structural platform that enables the design, formulation, and deployment of any AAL setting. It defines modules and connections in an abstract way, making them a comprehensive, functional, and systematic architecture.

E. *UniversAAL*

The UniversAAL system is an open platform designed to promote the development, delivery, and implementation of technology solutions towards assisted living environments. This system is utilized to encourage end-users (i.e., supported individuals, their parents, and communities), AAL-responsible authorities, and organizations involved in the creation and implementation of AAL services. It comprises of a wide variety of tools (some are applications and others are models/architectures) targeted at these various classes. Services are divided into three major groups: runtime assistance, production assistance, and community assistance. UniversAAL is regarded to be one of the most comprehensive RMs to date [29,30]. It represents an interpretation of AAL structures at the maximum level of abstraction [31], utilizing as few terms as possible. It also reflects the AAL domain description, the AAL space review, the types of technology included, and several other principles.

F. *PERSONA*

PERSONA is a service platform for AAL environments; it is designed to promote the incremental creation of AAL areas focused on a compact foundation. AAL spaces are modelled as accessible distributed systems in PERSONA. The system depends on the administrative re-configuration of the platform elements, including the Situation Reasoner, the Dialog Manager, and the Service Orchestra, to provide aggregated benefits [32].

G. *SOPRANO*

SOPRANO is a highly configurable, open AAL framework for senior citizens focused on semantic agreements. This serves as a facilitating artifact among the different modules of the process by establishing a uniform integrated terminology for various layers of abstraction. The architecture of SOPRANO offers exclusively predetermined contract-based interfaces for various stakeholders centered on structured ontology [33]. Ontology serves as a facilitating artifact between the modules of the different structure by providing a shared interconnected language for different levels of abstraction. The SOPRANO initiative has been an important guide for several other initiatives, such as OpenAAL [34], which acts as a comprehensive ontology for AAL symmetrical models.

H. OpenAAL

OpenAAL is a collaborative open-source program of the FZI Research Center for Information Technologies, Friedrich-Schiller-University of Jena, and CAS Software AG. It provides a scalable and efficient interface for AAL situations and is focused on the research findings of many German and foreign initiatives, such as the SOPRANO2 Combined Project. The OpenAAL framework allows for easy deployment, configuration, and scenario-dependent provision of versatile, context-aware, and customized IT-based functions [29].

I. MPOWER

MPOWER is an AAL initiative with the goal of creating and maintaining AAL frameworks based on trends, service-oriented frameworks, online services, and Extensible Stylesheet Language (XSDL) transformations [35]. It emphasizes compatibility among services, extensively in the areas of AAL.

3.2. Step 2: Fuzzy AHP

Saaty suggested the Analytic Hierarchy Process (AHP) method in 1990. All numerical and contextual considerations are integrated into AHP in the decision-making procedure. Because the highly classified scale of 1 to 9 is often used in the AHP process, this method is commonly criticized for not integrating uncertainty into the decision-making procedure. The fuzzy AHP approach has also been used in many fields to overcome multi-criteria challenges. Haq and Kannan [36] used this approach to pick the best supplier in the supply chain. This was also used by Huang et al. [37] for the evaluation of R&D projects. Fuzzy set theory is essentially a type of classical set theory. It is based on an adjacency matrix, and assigns a rank between one and ten. If the symbol is a fuzzy package, a tilde (i.e., $\sim$) is placed over it. A fuzzy activity is defined by (l, m, u) whereby 'l' is the lowest number, 'm' is the most probable value, and 'u' is the maximum priority [19].

3.3. Step 3: Ranking with TOPSIS

On the basis of the results, alternatives are rated using TOPSIS. In this process, the following types of parameters or characteristics are regarded:

- Domain and context
- Goals
- Design and infrastructure
- Quality attributes
- Critical attributes

In this analysis, different kinds of alternatives are evaluated as follows:

- Negative ideal solution
- Positive ideal solution

TOPSIS is focused on the choice of the most appropriate alternative or initiative that is the furthest from the negative ideal solution and the nearest to the positive ideal solution. The positive and negative ideal solutions are those with the maximum and minimum benefits, respectively.

4. Results

This section addresses the various quantitative measurements of the integrated fuzzy deployment of the AHP-TOPSIS symmetrical method. To achieve this goal, in our research study, we used the combined fuzzy AHP-TOPSIS approach, a well-established and verified decision-making technique. This methodology is designed to evaluate different Ambient Assisted Living system's symmetrical models based on their impact assessment in the current information technology era. To create a more compelling result, we took recommendations from 79 experts with diverse technologies and academic abilities.

To evaluate the different Ambient Assisted Living system's symmetrical models from a user perspective, five Level-1 parameters, namely domain and context, goals, design and infrastructure, quality attributes, and critical attributes, were defined respectively as LC1, LC2, LC3, LC4 and LC5. Further sub-parameters for domain and context were usability, independency, and purpose, defined respectively as LC11, LC12 and LC13, LC14 and LC5. Goals sub-parameters were standardization and facilitation, defined respectively as LC21 and LC22. Sub-parameters for design and infrastructure were components and connections, interfaces, policies and guidelines, details, concretization, formality, data flows, and cost, defined respectively as LC31, LC32, LC33, LC34, LC35, LC36, LC37 and LC38. Quality attributes sub-parameters were interoperability, scalability, confidentiality, maintainability, privacy, trustability, and security, defined respectively as LC41, LC42, LC43, LC44, LC45, LC46 and LC47. Critical attributes sub-parameters were reliability, integrity, performance, availability, and safety, defined respectively as LC51, LC52, LC53, LC54 and LC55. Different alternatives to the Ambient Assisted Living system's symmetrical model were AmIRA, Continua, FeelGood, RAFAALS, UniversAAL, PERSONA, SOPRANO, OpenAAL, and MPOWER denoted by AT1, AT2, AT3, AT4, AT5, AT6, AT7, AT8 and AT9 respectively. The local criteria and sub-criteria weights were calculated using pair-wise comparison matrices.

The pair-wise comparative matrix for the level 1 factor was created, as shown in Table 1. The compound pair-wise relative matrixes for the hierarchical diagram of level 2 are also specified in Tables 2–6. Table 7 shows the defuzzification matrix with alpha cut method and local weights. Tables 8–12 show aggregated pair-wise comparison matrixes at level 2 for domain and context, goals, design and infrastructure, quality attributes, and critical attributes, respectively. To be more specific, integration was executed to quantify the element weights of each point. In addition, with the support of the hierarchical structure, Table 13 and Figure 4 represent the overall weights and ranking of methods.

Table 1. Fuzzy pair-wise comparison matrix at level 1.

Level 1	LC1	LC2	LC3	LC4	LC5
LC1	1.00000, 1.00000, 1.00000	1.87222, 2.52710, 3.20315	1.46140, 1.68142, 1.97431	1.44161, 2.43185, 3.38615	0.46177, 0.57214, 0.78451
LC2	-	1.00000, 1.00000, 1.00000	0.60183, 0.77154, 1.02165	0.77108, 0.9504, 1.21361	0.16130, 0.19513, 0.24917
LC3	-	-	1.00000, 1.00000, 1.00000	0.71694, 1.01502, 1.35153	0.20186, 0.24162, 0.31117
LC4	-	-	-	1.00000, 1.00000, 1.00000	0.19516, 0.22813, 0.21903
LC5	-	-	-	-	1.00000, 1.00000, 1.00000

Table 2. Fuzzy aggregated pair-wise comparison matrix at level 2 for domain and context.

Level 2 for 1	LC11	LC12	LC13
LC11	1.00000, 1.00000, 1.00000	0.68918, 0.88160, 1.10012	0.22515, 0.27612, 0.35714
LC12	-	1.00000, 1.00000, 1.00000	0.30151, 0.38912, 0.56091
LC13	-	-	1.00000, 1.00000, 1.00000

Table 3. Fuzzy aggregated pair-wise comparison matrix at level 2 for goals.

Level 2 for 2	LC21	LC22
LC21	1.00000, 1.00000, 1.00000	0.65751, 1.16531, 1.68831
LC22	-	1.00000, 1.00000, 1.00000

Table 4. Fuzzy aggregated pair-wise comparison matrix at level 2 for design and infrastructure.

	LC31	LC32	LC33	LC34	LC35	LC36	LC37	LC38
LC31	1.00000, 1.00000, 1.00000	1.00100, 1.51157, 1.93311	0.48916, 0.63712, 1.00010	0.41152, 0.57413, 1.00001	0.22115, 0.28171, 0.41152	0.31146, 0.46110, 0.87015	0.65175, 1.16513, 1.68813	0.24144, 0.32318, 0.48011
LC32	-	1.00000, 1.00000, 1.00000	0.57413, 0.66517, 0.80221	0.30319, 0.39316, 0.56611	0.26719, 0.35211, 0.51716	0.16613, 0.19619, 0.25311	0.39310, 0.57413, 1.05614	0.16912, 0.20716, 0.27519
LC33	-	-	1.00000, 1.00000, 1.00000	1.0000, 1.3195, 1.5518	0.3009, 0.4352, 0.8027	0.80127, 0.87015, 1.00010	1.26119, 1.82510, 2.43314	0.17218, 0.20911, 0.26481
LC34	-	-	-	1.00000, 1.00000, 1.00000	0.53186, 0.91143, 1.58316	0.60813, 1.05192, 1.68219	0.75103, 1.34165, 1.96111	0.67910, 0.74819, 0.87105
LC35	-	-	-	-	1.00000, 1.00000, 1.00000	0.41152, 0.63712, 1.17191	0.94165, 1.10195, 1.24157	0.25100, 0.33100, 0.50100
LC36	-	-	-	-	-	1.00000, 1.00000, 1.00000	1.88181, 2.55108, 3.16197	0.80127, 1.03152, 1.31160
LC37	-	-	-	-	-	-	1.00000, 1.00000, 1.00000	0.21136, 0.21575, 0.31195
LC38	-	-	-	-	-	-	-	1.00000, 1.00000, 1.00000

Table 5. Fuzzy aggregated pair-wise comparison matrix at level 2 for quality attributes.

	LC41	LC42	LC43	LC44	LC45	LC46	LC47
LC41	1.00000, 1.00000, 1.00000	0.31127, 0.43195, 0.62152	0.87313, 0.90112, 0.94165	0.22161, 0.29128, 0.41166	0.16163, 0.19169, 0.21531	0.22161, 0.29128, 0.41166	0.16163, 0.19169, 0.25311
LC42	-	1.00000, 1.00000, 1.00000	0.5743, 0.6657, 0.8022	0.3039, 0.3936, 0.5661	0.8027, 0.8705, 1.0000	0.9465, 1.1095, 1.2457	1.26119, 1.82510, 2.43314
LC43	-	-	1.00000, 1.00000, 1.00000	1.00010, 1.31915, 1.55118	0.60813, 1.05912, 1.68219	1.88811, 2.55018, 3.16917	0.75103, 1.34165, 1.96111
LC44	-	-	-	1.00000, 1.00000, 1.00000	0.53816, 0.91413, 1.58316	0.60813, 1.05912, 1.68219	0.75013, 1.34615, 1.96111
LC45	-	-	-	-	1.00000, 1.00000, 1.00000	0.41152, 0.63172, 1.17191	0.94615, 1.10915, 1.24517
LC46	-	-	-	-	-	1.00000, 1.00000, 1.00000	1.88181, 2.55108, 3.16197
LC47	-	-	-	-	-	-	1.00000, 1.00000, 1.00000

Table 6. Fuzzy aggregated pair-wise comparison matrix at level 2 for critical attributes.

	LC51	LC52	LC53	LC54	LC55
LC51	1.00000, 1.00000, 1.00000	0.97110, 1.24175, 1.60194	1.05912, 1.58419, 2.22016	0.77313, 1.01118, 1.28181	0.761112, 0.912110, 1.096115
LC52	-	1.00000, 1.00000, 1.00000	0.63512, 0.91143, 1.34310	0.42713, 0.63315, 0.96610	0.34716, 0.49010, 0.87314
LC53	-	-	1.00000, 1.00000, 1.00000	0.51416, 0.65715, 0.78146	0.52113, 0.65917, 0.91191
LC54	-	-	-	1.00000, 1.00000, 1.00000	0.55612, 0.64148, 0.81122
LC55	-	-	-	-	1.00000, 1.00000, 1.00000

Table 7. Defuzzification matrix with alpha cut method and local weights.

Level 1	LC1	LC2	LC3	LC4	LC5	Weights
LC1	1.00000	2.55144	1.71017	2.42174	0.59193	0.240000
LC2	0.39115	1.00000	0.79164	0.97169	0.20173	0.095200
LC3	0.58176	1.25516	1.00000	1.05163	0.25132	0.120000
LC4	0.41120	1.02136	0.94167	1.00000	0.23157	0.103200
LC5	1.66186	4.82139	3.94195	4.21427	1.00000	0.441600
			CR = 0.00250254			

Table 8. Aggregated pair-wise comparison matrix at level 2 for domain and context.

Level 2 for 1	LC11	LC12	LC13	Weights
LC11	1.00000	0.81905	0.28139	0.183200
LC12	1.12130	1.00000	0.41111	0.223900
LC13	3.52124	2.43125	1.00000	0.592900
		CR = 0.0062		

Table 9. Aggregated pair-wise comparison matrix at level 2 for goals.

Level 2 for 2	LC21	LC22	Weights
LC21	1.00000	1.16911	0.5391000
LC22	0.85154	1.00000	0.4611000
		C.R. = 0.001	

Table 10. Aggregated pair-wise comparison matrix at level 2 for design and infrastructure.

	LC31	LC32	LC33	LC34	LC35	LC36	LC37	LC38	Weights
LC31	1.00000	1.409012	0.69001	0.60410	0.30002	0.52060	1.10069	0.34300	0.073300
LC32	0.67006	1.00000	0.67700	0.40143	0.37204	0.20033	0.64095	0.21051	0.049700
LC33	1.44070	1.47071	1.00000	1.29077	0.49305	0.85200	1.83064	0.21040	0.103100
LC34	1.56000	2.41307	0.77006	1.00000	0.96036	1.10024	1.35011	0.73190	0.127100
LC35	3.30306	2.68503	2.02603	1.03078	1.00000	0.71072	1.10028	0.43500	0.141400
LC36	1.89802	4.91808	1.17307	0.90071	1.39043	1.00000	2.38052	1.00473	0.172900
LC37	0.85540	1.53907	0.54405	0.74001	0.90679	0.41902	1.00000	0.26021	0.076000
LC38	2.91054	4.64900	4.67029	1.36631	2.29089	0.95048	3.81053	1.00000	0.256500
C.R. = 0.0333									

Table 11. Aggregated pair-wise comparison matrix at level 2 for quality attributes.

	LC41	LC42	LC43	LC44	LC45	LC46	LC47	Weights
LC41	1.00000	1.4912	0.69010	0.64010	0.30207	0.52608	1.16901	0.098700
LC42	0.67006	1.00000	0.6770	0.41403	0.37024	0.20033	0.64905	0.120700
LC43	1.44700	1.47701	1.00000	1.29077	0.49305	0.85200	1.83064	0.119600
LC44	1.56000	2.41370	0.77006	1.00000	0.96306	1.10204	1.35101	0.166500
LC45	3.30360	2.68503	2.02063	1.03708	1.00000	0.71702	1.10028	0.178500
LC46	1.89082	4.91088	1.17037	0.90071	1.39043	1.00000	2.38052	0.116000
LC47	0.85054	1.53907	0.54045	0.74001	0.90679	0.41925	1.00000	0.200000
C.R. = 0.03548								

Table 12. Aggregated pair-wise comparison matrix at level 2 for critical attributes.

	LC51	LC52	LC53	LC54	LC55	Weights
LC51	1.00000	1.26890	1.61240	1.02103	0.92004	0.221600
LC52	0.78081	1.00000	1.26903	0.66501	0.55003	0.159600
LC53	0.62002	0.78708	1.00000	0.65306	0.69000	0.144600
LC54	0.97901	1.50305	1.53000	1.00000	0.66405	0.211500
LC55	1.08605	1.81702	1.44903	1.50409	1.00000	0.262700
C.R. = 0.0069						

Table 13. Overall weights and ranking of methods.

Level 1	Local Weights	Level 2	Local Weights	Global Weights	Normalized Weights
LC1	0.240000	LC11	0.183200	0.043968	0.045086
		LC12	0.223900	0.053736	0.055103
		LC13	0.592900	0.142296	0.145915
LC2	0.095200	LC21	0.539000	0.051313	0.052618
		LC22	0.461000	0.043887	0.045003

Table 13. Cont.

Level 1	Local Weights	Level 2	Local Weights	Global Weights	Normalized Weights
LC3	0.120000	LC31	0.073300	0.006978	0.007155
		LC32	0.049700	0.004732	0.004852
		LC33	0.103100	0.009815	0.010065
		LC34	0.127100	0.012099	0.012407
		LC35	0.141400	0.013461	0.013803
		LC36	0.172900	0.016460	0.016879
		LC37	0.076000	0.007235	0.007419
		LC38	0.256500	0.024419	0.02504
LC4	0.103200	LC41	0.098700	0.010186	0.010445
		LC42	0.120700	0.012456	0.012773
		LC43	0.119600	0.012343	0.012657
		LC44	0.166500	0.017183	0.01762
		LC45	0.178500	0.018421	0.018889
		LC46	0.116000	0.011971	0.012275
		LC47	0.200000	0.020640	0.021165
LC5	0.441600	LC51	0.221600	0.097859	0.100348
		LC52	0.159600	0.070479	0.072271
		LC53	0.144600	0.063855	0.065479
		LC54	0.211500	0.093398	0.095773
		LC55	0.262700	0.116008	0.118958

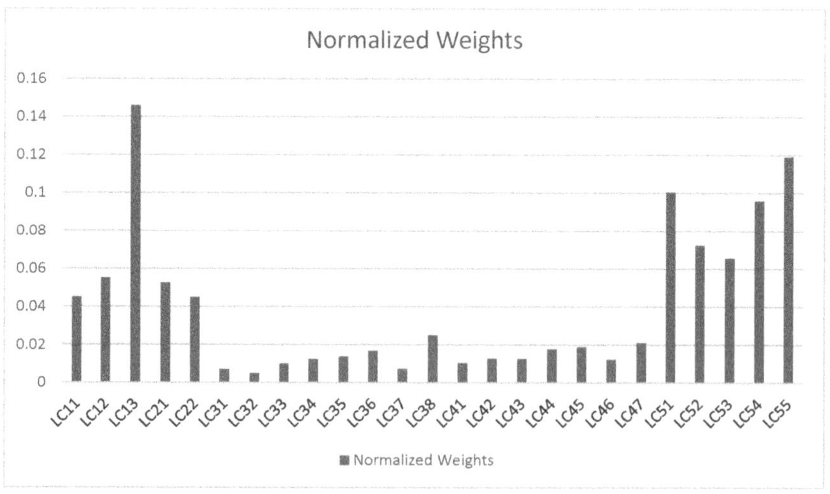

Figure 4. Graphical representation of overall weights.

Table 14 shows the subjective cognition results of evaluators in linguistic terms. Table 15 presents the normalized fuzzy decision matrix. Table 16 presents the weighted normalized fuzzy decision matrix. Finally, Table 17 and Figure 5 shows the closeness coefficients to the aspired level among the different alternatives.

Table 14. Subjective cognition results of evaluators in linguistic terms.

	AT1	AT2	AT3	AT4	AT5	AT6	AT7	AT8	AT9
LC11	5.120, 7.140, 8.720	3.150, 5.150, 6.910	2.820, 4.640, 6.640	1.550, 3.180, 5.180	5.120, 7.140, 8.720	3.150, 5.150, 6.910	2.820, 4.640, 6.640	1.550, 3.180, 5.180	1.450, 3.180, 5.180
LC12	4.280, 6.370, 8.370	2.820, 4.640, 6.640	1.550, 3.180, 5.180	5.120, 7.140, 8.720	3.150, 5.150, 6.910	2.450, 4.450, 6.450	2.910, 4.640, 6.550	1.450, 3.000, 4.910	1.180, 2.820, 4.820
LC13	4.270, 6.270, 8.140	2.910, 4.640, 6.550	1.450, 3.000, 4.910	4.280, 6.370, 8.370	2.820, 4.640, 6.640	1.550, 3.180, 5.180	5.120, 7.140, 8.720	3.150, 5.150, 6.910	2.450, 4.450, 6.450
LC21	5.360, 7.360, 9.120	5.120, 7.140, 8.720	3.150, 5.150, 6.910	2.820, 4.640, 6.640	2.910, 4.640, 6.550	1.450, 3.000, 4.910	4.280, 6.370, 8.370	2.450, 4.450, 6.450	2.820, 4.640, 6.640
LC22	4.640, 6.640, 8.550	2.820, 4.640, 6.640	1.550, 3.180, 5.180	5.120, 7.140, 8.720	3.150, 5.150, 6.910	3.150, 5.150, 6.910	2.820, 4.640, 6.640	5.360, 7.360, 9.120	2.450, 4.450, 6.450
LC31	3.120, 5.000, 7.140	2.910, 4.640, 6.550	1.450, 3.000, 4.910	4.280, 6.370, 8.370	2.450, 4.450, 6.450	2.450, 4.450, 6.450	2.910, 4.640, 6.550	4.640, 6.640, 8.550	2.450, 4.450, 6.450
LC32	4.280, 6.370, 8.370	5.120, 7.140, 8.720	3.150, 5.150, 6.910	2.820, 4.640, 6.640	5.360, 7.360, 9.120	2.820, 4.640, 6.640	2.820, 4.640, 6.640	5.360, 7.360, 9.120	1.180, 2.820, 4.820
LC33	4.280, 6.370, 8.370	4.280, 6.370, 8.370	2.450, 4.450, 6.450	2.910, 4.640, 6.550	4.640, 6.640, 8.550	1.820, 3.730, 5.730	2.820, 4.640, 6.640	5.360, 7.360, 9.120	1.180, 2.820, 4.820
LC34	4.270, 6.270, 8.140	2.180, 4.090, 6.140	2.820, 4.640, 6.640	2.820, 4.640, 6.640	5.360, 7.360, 9.120	1.450, 3.000, 4.910	4.280, 6.370, 8.370	2.450, 4.450, 6.450	2.450, 4.450, 6.450
LC35	4.280, 6.370, 8.370	3.550, 5.550, 7.450	1.820, 3.730, 5.730	2.820, 4.640, 6.640	5.360, 7.360, 9.120	1.450, 3.000, 4.910	4.280, 6.370, 8.370	2.450, 4.450, 6.450	1.180, 2.820, 4.820
LC36	4.270, 6.270, 8.140	2.910, 4.640, 6.550	1.450, 3.000, 4.910	4.280, 6.370, 8.370	2.450, 4.450, 6.450	3.150, 5.150, 6.910	2.820, 4.640, 6.640	5.360, 7.360, 9.120	2.450, 4.450, 6.450
LC37	5.360, 7.360, 9.120	2.910, 4.640, 6.550	1.450, 3.000, 4.910	4.280, 6.370, 8.370	2.450, 4.450, 6.450	1.450, 3.000, 4.910	4.280, 6.370, 8.370	2.450, 4.450, 6.450	2.820, 4.640, 6.640
LC38	4.280, 6.370, 8.370	5.120, 7.140, 8.720	3.150, 5.150, 6.910	2.820, 4.640, 6.640	5.360, 7.360, 9.120	3.150, 5.150, 6.910	2.820, 4.640, 6.640	5.360, 7.360, 9.120	1.180, 2.820, 4.820
LC41	4.270, 6.270, 8.140	2.910, 4.640, 6.550	1.450, 3.000, 4.910	4.280, 6.370, 8.370	2.450, 4.450, 6.450	2.450, 4.450, 6.450	2.910, 4.640, 6.550	4.640, 6.640, 8.550	2.450, 4.450, 6.450
LC42	5.360, 7.360, 9.120	5.120, 7.140, 8.720	3.150, 5.150, 6.910	2.820, 4.640, 6.640	5.360, 7.360, 9.120	1.450, 3.000, 4.910	4.280, 6.370, 8.370	2.450, 4.450, 6.450	2.820, 4.640, 6.640
LC43	4.280, 6.370, 8.370	4.280, 6.370, 8.370	2.450, 4.450, 6.450	2.910, 4.640, 6.550	4.640, 6.640, 8.550	3.150, 5.150, 6.910	2.820, 4.640, 6.640	5.360, 7.360, 9.120	1.180, 2.820, 4.820
LC44	4.270, 6.270, 8.140	2.910, 4.640, 6.550	1.450, 3.000, 4.910	4.280, 6.370, 8.370	2.450, 4.450, 6.450	2.450, 4.450, 6.450	2.910, 4.640, 6.550	4.640, 6.640, 8.550	2.450, 4.450, 6.450

Table 14. *Cont.*

	AT1	AT2	AT3	AT4	AT5	AT6	AT7	AT8	AT9
LC45	5.360, 7.360, 9.120	5.120, 7.140, 8.720	3.150, 5.150, 6.910	2.820, 4.640, 6.640	5.360, 7.360, 9.120	1.450, 3.000, 4.910	4.280, 6.370, 8.370	2.450, 4.450, 6.450	2.820, 4.640, 6.640
LC46	4.280, 6.370, 8.370	4.280, 6.370, 8.370	2.450, 4.450, 6.450	2.910, 4.640, 6.550	4.640, 6.640, 8.550	3.150, 5.150, 6.910	2.820, 4.640, 6.640	5.360, 7.360, 9.120	1.180, 2.820, 4.820
LC47	4.270, 6.270, 8.140	2.910, 4.640, 6.550	1.450, 3.000, 4.910	4.280, 6.370, 8.370	2.450, 4.450, 6.450	2.450, 4.450, 6.450	2.910, 4.640, 6.550	4.640, 6.640, 8.550	2.450, 4.450, 6.450
LC51	5.360, 7.360, 9.120	5.120, 7.140, 8.720	3.150, 5.150, 6.910	2.820, 4.640, 6.640	5.360, 7.360, 9.120	1.450, 3.000, 4.910	4.280, 6.370, 8.370	2.450, 4.450, 6.450	2.820, 4.640, 6.640
LC52	4.640, 6.640, 8.550	4.280, 6.370, 8.370	2.450, 4.450, 6.450	2.910, 4.640, 6.550	4.640, 6.640, 8.550	3.150, 5.150, 6.910	2.820, 4.640, 6.640	5.360, 7.360, 9.120	2.450, 4.450, 6.450
LC53	3.120, 5.000, 7.140	2.910, 4.640, 6.550	1.450, 3.000, 4.910	4.280, 6.370, 8.370	2.450, 4.450, 6.450	2.450, 4.450, 6.450	2.910, 4.640, 6.550	4.640, 6.640, 8.550	2.450, 4.450, 6.450
LC54	5.360, 7.360, 9.120	5.120, 7.140, 8.720	3.150, 5.150, 6.910	2.820, 4.640, 6.640	5.360, 7.360, 9.120	2.820, 4.640, 6.640	2.820, 4.640, 6.640	5.360, 7.360, 9.120	2.820, 4.640, 6.640
LC55	4.640, 6.640, 8.550	4.280, 6.370, 8.370	2.450, 4.450, 6.450	2.910, 4.640, 6.550	4.640, 6.640, 8.550	5.360, 7.360, 9.120	3.730, 5.730, 7.550	4.280, 6.370, 8.370	2.450, 4.450, 6.450

Table 15. The normalized fuzzy decision matrix.

	AT1	AT2	AT3	AT4	AT5	AT6	AT7	AT8	AT9
LC11	0.460, 0.690, 0.910	0.320, 0.580, 0.850	0.390, 0.620, 0.870	0.460, 0.690, 0.910	0.320, 0.580, 0.850	0.390, 0.620, 0.870	0.210, 0.450, 0.730	0.180, 0.430, 0.74 0	0.210, 0.450, 0.730
LC12	0.460, 0.680, 0.890	0.370, 0.630, 0.900	0.420, 0.690, 0.950	0.460, 0.680, 0.890	0.370, 0.630, 0.900	0.420, 0.690, 0.950	0.210, 0.460, 0.730	0.120, 0.350, 0.660	0.300, 0.530, 0.790
LC13	0.560, 0.780, 0.950	0.410, 0.680, 0.910	0.370, 0.620, 0.890	0.560, 0.780, 0.950	0.410, 0.680, 0.910	0.370, 0.620, 0.890	0.230, 0.470, 0.780	0.220, 0.490, 0.800	0.260, 0.470, 0.720
LC21	0.460, 0.690, 0.910	0.460, 0.690, 0.910	0.320, 0.580, 0.850	0.390, 0.620, 0.870	0.460, 0.690, 0.910	0.320, 0.580, 0.850	0.390, 0.620, 0.870	0.210, 0.450, 0.730	0.180, 0.430, 0.74 0
LC22	0.460, 0.680, 0.890	0.460, 0.680, 0.890	0.370, 0.630, 0.900	0.420, 0.690, 0.950	0.460, 0.680, 0.890	0.370, 0.630, 0.900	0.420, 0.690, 0.950	0.210, 0.460, 0.730	0.120, 0.350, 0.660
LC31	0.460, 0.690, 0.910	0.320, 0.580, 0.850	0.390, 0.620, 0.870	0.210, 0.450, 0.730	0.180, 0.430, 0.74 0	0.410, 0.680, 0.910	0.370, 0.620, 0.890	0.230, 0.470, 0.780	0.220, 0.490, 0.800
LC32	0.460, 0.680, 0.890	0.370, 0.630, 0.900	0.420, 0.690, 0.950	0.210, 0.460, 0.730	0.460, 0.690, 0.910	0.320, 0.580, 0.850	0.390, 0.620, 0.870	0.210, 0.450, 0.730	0.180, 0.430, 0.740

Table 15. *Cont.*

	AT1	AT2	AT3	AT4	AT5	AT6	AT7	AT8	AT9
LC33	0.560, 0.780, 0.950	0.410, 0.680, 0.910	0.370, 0.620, 0.890	0.230, 0.470, 0.780	0.460, 0.680, 0.890	0.370, 0.630, 0.900	0.420, 0.690, 0.950	0.210, 0.460, 0.730	0.120, 0.350, 0.660
LC34	0.460, 0.690, 0.910	0.320, 0.580, 0.850	0.390, 0.620, 0.870	0.210, 0.450, 0.730	0.560, 0.780, 0.950	0.410, 0.680, 0.910	0.370, 0.620, 0.890	0.230, 0.470, 0.780	0.220, 0.490, 0.800
LC35	0.460, 0.680, 0.890	0.370, 0.630, 0.900	0.420, 0.690, 0.950	0.210, 0.460, 0.730	0.460, 0.690, 0.910	0.320, 0.580, 0.850	0.390, 0.620, 0.870	0.210, 0.450, 0.730	0.180, 0.430, 0.74 0
LC36	0.460, 0.690, 0.910	0.320, 0.580, 0.850	0.390, 0.620, 0.870	0.210, 0.450, 0.730	0.180, 0.430, 0.74 0	0.460, 0.690, 0.910	0.320, 0.580, 0.850	0.390, 0.620, 0.870	0.210, 0.450, 0.730
LC37	0.460, 0.680, 0.890	0.370, 0.630, 0.900	0.420, 0.690, 0.950	0.210, 0.460, 0.730	0.120, 0.350, 0.660	0.460, 0.680, 0.890	0.370, 0.630, 0.900	0.420, 0.690, 0.950	0.210, 0.460, 0.730
LC38	0.560, 0.780, 0.950	0.410, 0.680, 0.910	0.370, 0.620, 0.890	0.230, 0.470, 0.780	0.220, 0.490, 0.800	0.560, 0.780, 0.950	0.410, 0.680, 0.910	0.370, 0.620, 0.890	0.230, 0.470, 0.780
LC41	0.460, 0.690, 0.910	0.320, 0.580, 0.850	0.390, 0.620, 0.870	0.210, 0.450, 0.730	0.180, 0.430, 0.74 0	0.460, 0.690, 0.910	0.320, 0.580, 0.850	0.390, 0.620, 0.870	0.210, 0.450, 0.730
LC42	0.460, 0.680, 0.890	0.370, 0.630, 0.900	0.420, 0.690, 0.950	0.210, 0.460, 0.730	0.120, 0.350, 0.660	0.460, 0.680, 0.890	0.370, 0.630, 0.900	0.420, 0.690, 0.950	0.210, 0.460, 0.730
LC43	0.230, 0.470, 0.780	0.220, 0.490, 0.800	0.460, 0.690, 0.910	0.320, 0.580, 0.850	0.390, 0.620, 0.870	0.230, 0.470, 0.780	0.220, 0.490, 0.800	0.460, 0.690, 0.910	0.320, 0.580, 0.850
LC44	0.560, 0.780, 0.950	0.410, 0.680, 0.910	0.370, 0.620, 0.890	0.230, 0.470, 0.780	0.220, 0.490, 0.800	0.560, 0.780, 0.950	0.410, 0.680, 0.910	0.370, 0.620, 0.890	0.230, 0.470, 0.780
LC45	0.460, 0.690, 0.910	0.320, 0.580, 0.850	0.390, 0.620, 0.870	0.210, 0.450, 0.730	0.180, 0.430, 0.74 0	0.460, 0.690, 0.910	0.320, 0.580, 0.850	0.390, 0.620, 0.870	0.210, 0.450, 0.730
LC46	0.460, 0.690, 0.910	0.320, 0.580, 0.850	0.390, 0.620, 0.870	0.210, 0.450, 0.730	0.460, 0.690, 0.910	0.320, 0.580, 0.850	0.390, 0.620, 0.870	0.210, 0.450, 0.730	0.180, 0.430, 0.74 0
LC47	0.460, 0.680, 0.890	0.370, 0.630, 0.900	0.420, 0.690, 0.950	0.210, 0.460, 0.730	0.460, 0.680, 0.890	0.370, 0.630, 0.900	0.420, 0.690, 0.950	0.210, 0.460, 0.730	0.120, 0.350, 0.660
LC51	0.560, 0.780, 0.950	0.410, 0.680, 0.910	0.370, 0.620, 0.890	0.230, 0.470, 0.780	0.560, 0.780, 0.950	0.410, 0.680, 0.910	0.370, 0.620, 0.890	0.230, 0.470, 0.780	0.220, 0.490, 0.800
LC52	0.460, 0.690, 0.910	0.320, 0.580, 0.850	0.390, 0.620, 0.870	0.210, 0.450, 0.730	0.460, 0.690, 0.910	0.320, 0.580, 0.850	0.390, 0.620, 0.870	0.210, 0.450, 0.730	0.180, 0.430, 0.740
LC53	0.460, 0.680, 0.890	0.370, 0.630, 0.900	0.420, 0.690, 0.950	0.210, 0.460, 0.730	0.460, 0.680, 0.890	0.370, 0.630, 0.900	0.420, 0.690, 0.950	0.210, 0.460, 0.730	0.120, 0.350, 0.660

Table 15. *Cont.*

	AT1	AT2	AT3	AT4	AT5	AT6	AT7	AT8	AT9
LC54	0.230, 0.470, 0.780	0.220, 0.490, 0.800	0.460, 0.690, 0.910	0.320, 0.580, 0.850	0.230, 0.470, 0.780	0.220, 0.490, 0.800	0.460, 0.690, 0.910	0.320, 0.580, 0.850	0.390, 0.620, 0.870
LC55	0.560, 0.780, 0.950	0.410, 0.680, 0.910	0.370, 0.620, 0.890	0.230, 0.470, 0.780	0.560, 0.780, 0.950	0.410, 0.680, 0.910	0.370, 0.620, 0.890	0.230, 0.470, 0.780	0.220, 0.490, 0.800

Table 16. The weighted normalized fuzzy decision matrix.

	AT1	AT2	AT3	AT4	AT5	AT6	AT7	AT8	AT9
LC11	0.054, 0.120, 0.260	0.125, 0.155, 0.344	0.041, 0.095, 0.242	0.059, 0.121, 0.296	0.041, 0.100, 0.260	0.045, 0.098, 0.239	0.041, 0.095, 0.242	0.041, 0.100, 0.260	0.045, 0.098, 0.239
LC12	0.043, 0.096, 0.196	0.041, 0.095, 0.198	0.061, 0.121, 0.233	0.125, 0.155, 0.344	0.041, 0.095, 0.242	0.059, 0.121, 0.296	0.041, 0.100, 0.260	0.045, 0.098, 0.239	0.041, 0.095, 0.242
LC13	0.041, 0.095, 0.242	0.102, 0.137, 0.299	0.114, 0.144, 0.306	0.041, 0.095, 0.198	0.061, 0.121, 0.233	0.114, 0.144, 0.306	0.125, 0.155, 0.344	0.041, 0.095, 0.242	0.059, 0.121, 0.296
LC21	0.061, 0.121, 0.233	0.027, 0.080, 0.197	0.051, 0.104, 0.168	0.102, 0.137, 0.299	0.114, 0.144, 0.306	0.044, 0.088, 0.182	0.041, 0.095, 0.198	0.061, 0.121, 0.233	0.034, 0.091, 0.200
LC22	0.125, 0.155, 0.344	0.041, 0.095, 0.242	0.059, 0.121, 0.296	0.041, 0.100, 0.260	0.045, 0.098, 0.239	0.041, 0.095, 0.242	0.125, 0.155, 0.344	0.041, 0.095, 0.242	0.059, 0.121, 0.296
LC31	0.041, 0.095, 0.198	0.061, 0.121, 0.233	0.114, 0.144, 0.306	0.125, 0.155, 0.344	0.125, 0.155, 0.344	0.041, 0.095, 0.242	0.059, 0.121, 0.296	0.041, 0.100, 0.260	0.045, 0.098, 0.239
LC32	0.102, 0.137, 0.299	0.114, 0.144, 0.306	0.044, 0.088, 0.182	0.041, 0.095, 0.198	0.041, 0.095, 0.198	0.061, 0.121, 0.233	0.114, 0.144, 0.306	0.125, 0.155, 0.344	0.041, 0.095, 0.242
LC33	0.125, 0.155, 0.344	0.041, 0.095, 0.242	0.059, 0.121, 0.296	0.041, 0.100, 0.260	0.045, 0.098, 0.239	0.041, 0.095, 0.242	0.044, 0.088, 0.182	0.041, 0.095, 0.198	0.061, 0.121, 0.233
LC34	0.041, 0.095, 0.198	0.061, 0.121, 0.233	0.114, 0.144, 0.306	0.125, 0.155, 0.344	0.041, 0.095, 0.242	0.059, 0.121, 0.296	0.114, 0.144, 0.306	0.125, 0.155, 0.344	0.041, 0.095, 0.242
LC35	0.102, 0.137, 0.299	0.114, 0.144, 0.306	0.044, 0.088, 0.182	0.125, 0.155, 0.344	0.041, 0.095, 0.242	0.059, 0.121, 0.296	0.041, 0.100, 0.260	0.045, 0.098, 0.239	0.041, 0.095, 0.242
LC36	0.027, 0.080, 0.197	0.051, 0.104, 0.168	0.114, 0.144, 0.306	0.041, 0.095, 0.198	0.061, 0.121, 0.233	0.114, 0.144, 0.306	0.125, 0.155, 0.344	0.041, 0.095, 0.242	0.059, 0.121, 0.296
LC37	0.054, 0.120, 0.260	0.027, 0.080, 0.197	0.051, 0.104, 0.168	0.102, 0.137, 0.299	0.114, 0.144, 0.306	0.044, 0.088, 0.182	0.041, 0.095, 0.198	0.061, 0.121, 0.233	0.034, 0.091, 0.200
LC38	0.125, 0.155, 0.344	0.041, 0.095, 0.242	0.059, 0.121, 0.296	0.041, 0.100, 0.260	0.045, 0.098, 0.239	0.041, 0.095, 0.242	0.125, 0.155, 0.344	0.041, 0.095, 0.242	0.059, 0.121, 0.296

Table 16. *Cont.*

	AT1	AT2	AT3	AT4	AT5	AT6	AT7	AT8	AT9
LC41	0.041, 0.095, 0.198	0.061, 0.121, 0.233	0.114, 0.144, 0.306	0.125, 0.155, 0.344	0.125, 0.155, 0.344	0.041, 0.095, 0.242	0.059, 0.121, 0.296	0.041, 0.100, 0.260	0.045, 0.098, 0.239
LC42	0.102, 0.137, 0.299	0.114, 0.144, 0.306	0.044, 0.088, 0.182	0.041, 0.095, 0.198	0.041, 0.095, 0.198	0.061, 0.121, 0.233	0.114, 0.144, 0.306	0.125, 0.155, 0.344	0.041, 0.095, 0.242
LC43	0.027, 0.080, 0.197	0.051, 0.104, 0.168	0.114, 0.144, 0.306	0.125, 0.155, 0.344	0.102, 0.137, 0.299	0.114, 0.144, 0.306	0.044, 0.088, 0.182	0.041, 0.095, 0.198	0.061, 0.121, 0.233
LC44	0.043, 0.096, 0.196	0.017, 0.059, 0.152	0.036, 0.072, 0.162	0.044, 0.088, 0.182	0.027, 0.080, 0.197	0.051, 0.104, 0.168	0.114, 0.144, 0.306	0.125, 0.155, 0.344	0.041, 0.095, 0.242
LC45	0.125, 0.155, 0.344	0.041, 0.095, 0.242	0.059, 0.121, 0.296	0.041, 0.100, 0.260	0.125, 0.155, 0.344	0.041, 0.095, 0.242	0.059, 0.121, 0.296	0.041, 0.100, 0.260	0.045, 0.098, 0.239
LC46	0.041, 0.095, 0.198	0.061, 0.121, 0.233	0.114, 0.144, 0.306	0.125, 0.155, 0.344	0.041, 0.095, 0.198	0.061, 0.121, 0.233	0.114, 0.144, 0.306	0.125, 0.155, 0.344	0.041, 0.095, 0.242
LC47	0.102, 0.137, 0.299	0.114, 0.144, 0.306	0.044, 0.088, 0.182	0.041, 0.095, 0.198	0.102, 0.137, 0.299	0.114, 0.144, 0.306	0.044, 0.088, 0.182	0.041, 0.095, 0.198	0.061, 0.121, 0.233
LC51	0.125, 0.155, 0.344	0.041, 0.095, 0.242	0.059, 0.121, 0.296	0.041, 0.100, 0.260	0.125, 0.155, 0.344	0.041, 0.095, 0.242	0.059, 0.121, 0.296	0.041, 0.100, 0.260	0.045, 0.098, 0.239
LC52	0.041, 0.095, 0.198	0.061, 0.121, 0.233	0.114, 0.144, 0.306	0.125, 0.155, 0.344	0.041, 0.095, 0.198	0.061, 0.121, 0.233	0.114, 0.144, 0.306	0.125, 0.155, 0.344	0.041, 0.095, 0.242
LC53	0.102, 0.137, 0.299	0.114, 0.144, 0.306	0.044, 0.088, 0.182	0.041, 0.095, 0.198	0.102, 0.137, 0.299	0.114, 0.144, 0.306	0.044, 0.088, 0.182	0.041, 0.095, 0.198	0.061, 0.121, 0.233
LC54	0.027, 0.080, 0.197	0.051, 0.104, 0.168	0.114, 0.144, 0.306	0.125, 0.155, 0.344	0.027, 0.080, 0.197	0.051, 0.104, 0.168	0.114, 0.144, 0.306	0.125, 0.155, 0.344	0.041, 0.095, 0.242
LC55	0.043, 0.096, 0.196	0.017, 0.059, 0.152	0.036, 0.072, 0.162	0.044, 0.088, 0.182	0.041, 0.095, 0.198	0.061, 0.121, 0.233	0.034, 0.091, 0.200	0.032, 0.089, 0.200	0.063, 0.120, 0.233

Table 17. Closeness coefficients to the aspired level among the different alternatives.

Alternatives	d+i	d−i	Gap Degree of CC+i	Satisfaction Degree of CC−i
AT1	1.249124	1.333548	0.5164578	0.4832657
AT2	0.699528	0.840586	0.5476598	0.4532564
AT3	0.787654	1.484745	0.6535644	0.3452657
AT4	2.168547	1.484856	0.4125471	0.5923547
AT5	2.005654	1.536954	0.4332654	0.5652547
AT6	0.445476	0.392356	0.4645687	0.5345854
AT7	0.788574	1.484657	0.4125475	0.4522564
AT8	2.160256	1.536235	0.4332654	0.3452544
AT9	2.035657	0.397596	0.4645288	0.5912556

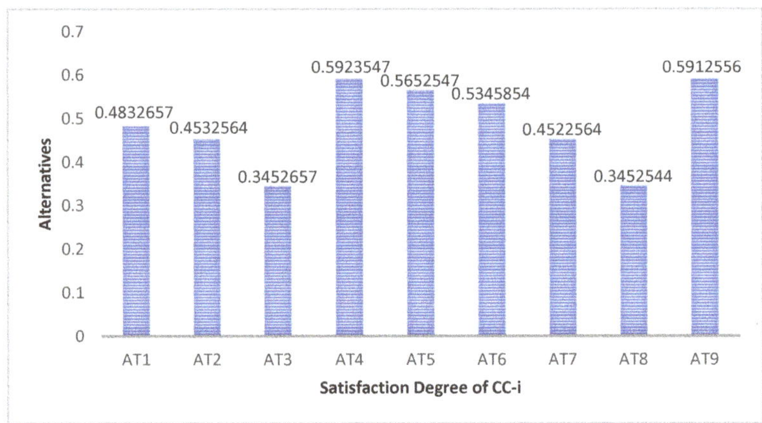

Figure 5. Graphical representation of closeness coefficients to the aspired level among the different alternatives.

Consequently, RAFAALS (denoted by AT4) was found to be best among nine comparative alternatives because it offers the best functionality with an efficiency score of 0.5923547 among the different Ambient Assisted Living system's symmetrical models. Alternative AT4 was followed by AT9, AT5, AT6, AT1, AT2, AT7, AT3, and AT8 with performance scores of 0.5912556, 0.5652547, 0.5345854, 0.4832657, 0.4532564, 0.4522564, 0.3452657 and 0.3452544, respectively.

5. Conclusions

AAL symmetrical models have progressed as an outcome of the emergence of global population ageing and the change of direction of technological advances. It lies at the intersection of technological innovation and age advancement. This multidisciplinary area of scientific research sees technology as a way of enhancing the lives of older people and promoting their involvement as involved members of society. This paper explores the principles of the Ambient Assisted Living area. We tested predefined categories to illustrate the key areas that need to change due to the advancement of AAL systems. We suggest that there is a lack of standardization in several of the available frameworks. This is the key deterrent in achieving the desired efficacy of the models. Maintenance and interoperability occur in all of the models. In some systems, there is a need for facility, interoperability, and independence. For a specific design model, the practitioners need to focus on a solid standard and develop a robust infrastructure. One of the crucial aspects of AAL systems is coping with data flow, which remains a largely unexplored domain and needs more dedicated research.

Author Contributions: W.A. performed formal analysis and investigation. M.T.J.A. conceived and designed the experiments, performed the computation work authored and reviewed drafts of the paper. A.A. (Abdullah Alharbi) contributed reagents/materials/analysis tools. H.A. performed the visualization, review and editing work. A.H.S. analyzed and reviewed the drafts. A.K.P. performed the review and editing work. A.A. (Alka Agrawal) analyzed the data, contributed reagents/materials/analysis tools, authored or reviewed drafts of the paper. R.A.K. conceived and designed the experiments, performed the experiments, contributed reagents/materials/analysis tools. All authors have read and agreed to the published version of the manuscript.

Funding: The project has been funded by Taif University, Kingdom of Saudi Arabia.

Institutional Review Board Statement: Not applicable.

Informed Consent Statement: Not applicable.

Data Availability Statement: Not applicable.

Acknowledgments: This research was supported by Taif University Researchers Supporting Project number (TURSP-2020/254), Taif University, Taif, Saudi Arabia.

Conflicts of Interest: The authors declare no conflict of interest.

References

1. Chuah, S.H.-W.; Rauschnabel, P.A.; Krey, N.; Nguyen, B.; Ramayah, T.; Lade, S. Wearable technologies: The role of usefulness and visibility in smartwatch adoption. *Comput. Hum. Behav.* **2016**, *65*, 276–284. [CrossRef]
2. National Research Council. *Preparing for an Aging World: The Case for Cross-National Research*; National Academies Press: Washington, DC, USA, 2001.
3. Sun, H.; De Florio, V.; Gui, N.; Blondia, C. Promises and challenges of ambient assisted living systems. In Proceedings of the 2009 Sixth International Conference on Information Technology: New Generations, Las Vegas, NV, USA, 27–29 April 2009; pp. 1201–1207.
4. Rashidi, P.; Mihailidis, A. A survey on ambient-assisted living tools for older adults. *IEEE J. Biomed. Health Inform.* **2012**, *17*, 579–590. [CrossRef]
5. Freitas, P.; Menezes, P.; Dias, J. Ambient Assisted Living—From Technology to Intervention. In *Ambient Assisted Living*; CRC Press: Boca Raton, FL, USA, 2015; pp. 384–421.
6. Stopczynski, A.; Stahlhut, C.; Larsen, J.E.; Petersen, M.K.; Hansen, L.K. The Smartphone Brain Scanner: A Portable Real-Time Neuroimaging System. *PLoS ONE* **2014**, *9*, e86733. [CrossRef] [PubMed]
7. Memon, M.; Wagner, S.R.; Pedersen, C.F.; Beevi, F.H.A.; Hansen, F.O. Ambient Assisted Living Healthcare Frameworks, Platforms, Standards, and Quality Attributes. *Sensors* **2014**, *14*, 4312–4341. [CrossRef]
8. Allen, N.B.; Nelson, B.W.; Brent, D.; Auerbach, R.P. Short-term prediction of suicidal thoughts and behaviors in adolescents: Can recent developments in technology and computational science provide a breakthrough? *J. Affect. Disord.* **2019**, *250*, 163–169. [CrossRef] [PubMed]
9. Abtoy, A.; Touhafi, A.; Tahiri, A. Ambient Assisted living system's models and architectures: A survey of the state of the art. *J. King Saud Univ. Comput. Inf. Sci.* **2020**, *32*, 1–10.
10. Ghodsypour, S.H.; O'Brien, C. A decision support system for supplier selection using an integrated analytic hierarchy process and linear programming. *Int. J. Prod. Econ.* **1998**, *56*, 199–212. [CrossRef]
11. Carney, D.J.; Wallnau, K.C. A basis for evaluation of commercial software. *Inf. Softw. Technol.* **1998**, *40*, 851–860. [CrossRef]
12. Hwang, C.L.; Yoon, K. *Multiple Attribute Decision Making*; Springer: Berlin/Heidelberg, Germany, 1981.
13. Dursun, M.; Karsak, E.E. A fuzzy MCDM approach for personnel selection. *Expert Syst. Appl.* **2010**, *37*, 4324–4330. [CrossRef]
14. Lin, Y.-C.; Wang, Y.-C.; Chen, T.-C.T.; Lin, H.-F. Evaluating the Suitability of a Smart Technology Application for Fall Detection Using a Fuzzy Collaborative Intelligence Approach. *Mathematics* **2019**, *7*, 1097. [CrossRef]
15. Samanlioglu, F.; Taskaya, Y.E.; Gulen, U.C.; Cokcan, O. A fuzzy AHP–TOPSIS-based group decision-making approach to IT personnel selection. *Int. J. Fuzzy Syst.* **2018**, *20*, 1576–1591. [CrossRef]
16. Anand, M.B.; Vinodh, S. Application of fuzzy AHP–TOPSIS for ranking additive manufacturing processes for mi-crofabrication. *Rapid Prototyp. J.* **2018**, *24*, 424–435. [CrossRef]
17. Nazam, M.; Xu, J.; Tao, Z.; Ahmad, J.; Hashim, M. A fuzzy AHP-TOPSIS framework for the risk assessment of green supply chain implementation in the textile industry. *Int. J. Supply Oper. Manag.* **2015**, *2*, 548–568.
18. Ansari, T.J.; Al-Zahrani, F.A.; Pandey, D.; Agrawal, A. A fuzzy TOPSIS based analysis toward selection of effective security requirements engineering approach for trustworthy healthcare software development. *BMC Med. Inf. Decis. Mak.* **2020**, *20*, 1–13. [CrossRef]
19. Kumar, R.; Alenezi, M.; Ansari; Gupta, B.; Agrawal, A.; Khan, R.; Prince Sultan University; Shri Ramswaroop Memorial University. Evaluating the Impact of Malware Analysis Techniques for Securing Web Applications through a Decision-Making Framework under Fuzzy Environment. *Int. J. Intell. Eng. Syst.* **2020**, *13*, 94–109. [CrossRef]
20. Alenezi, M.; Agrawal, A.; Kumar, R.; Khan, R.A. Evaluating Performance of Web Application Security Through a Fuzzy Based Hybrid Multi-Criteria Decision-Making Approach: Design Tactics Perspective. *IEEE Access* **2020**, *8*, 25543–25556. [CrossRef]
21. Coronato, A.; Paragliola, G. A structured approach for the designing of safe aal applications. *Expert Syst. Appl.* **2017**, *85*, 1–13. [CrossRef]
22. Koleva, P.; Tonchev, K.; Balabanov, G.; Manolova, A.; Poulkov, V. Challenges in designing and implementation of an effective Ambient Assisted Living system. In Proceedings of the 2015 12th International Conference on Telecommunication in Modern Satellite, Cable and Broadcasting Services (TELSIKS), Nis, Serbia, 14–17 October 2015; pp. 305–308.
23. Berger, M.; Fuchs, F.; Pirker, M. Ambient intelligence-from personal assistance to intelligent megacities. *Front. Artif. Intell. Appl.* **2007**, *164*, 21.
24. Augusto, J.C.; Shapiro, D. (Eds.) *Advances in Ambient Intelligence*; IOS Press: Amsterdam, The Netherlands, 2007; Volume 164.
25. Wartena, F.; Muskens, J.; Schmitt, L.; Petković, M. Continua: The reference architecture of a personal telehealth ecosystem. In Proceedings of the 12th IEEE International Conference on e-Health Networking, Applications and Services, Lyon, France, 1–3 July 2010; pp. 1–6.

26. Hietala, H.; Ikonen, V.; Korhonen, I.; Lahteenmaki, K.; Maksimainen, A.; Pakarinen, V.; Saranummi, N. *Feelgood-Ecosystem of Phr Based Products and Services*; Research Report VTT-R-07000–09; VTT Technical Research Centre of Finland: Tampere, Finland, 2009.
27. Anouar, A.; Abdellah, T.; Abderahim, T. A novel reference model for ambient assisted living systems' architectures. *Appl. Comput. Inform.* **2020**. [CrossRef]
28. Amina, E.M.; Anouar, A.; Touhafi, A.; Tahiri, A. Towards an SOA Architectural Model for AALPaas Design and Implimentation Challenges. *Int. J. Adv. Comput. Sci. Appl.* **2017**, *8*, 52–56.
29. Garcés, L.; Ampatzoglou, A.; Oquendo, F.; Nakagawa, E.Y. *Assessment of Reference Architectures and Reference Models for Ambient Assisted Living Systems: A Systematic Literature Review*; IGI Global: São Paulo, Brazil, 2017.
30. Salvi, D.; MontalvaColomer, J.B.; Arredondo, M.T.; Prazak-Aram, B.; Mayer, C. A framework for evaluating Ambient Assisted Living technologies and the experience of the universAAL project. *J. Ambient Intell. Smart Environ.* **2015**, *7*, 329–352. [CrossRef]
31. Ferro, E.; Girolami, M.; Salvi, D.; Mayer, C.; Gorman, J.; Grguric, A.; Stocklöw, C. The universaal platform for aal (ambient assisted living). *J. Intell. Syst.* **2015**, *24*, 301–319. [CrossRef]
32. Tazari, M.-R.; Furfari, F.; Ramos, J.-P.L.; Ferro, E. The PERSONA Service Platform for AAL Spaces. In *Handbook of Ambient Intelligence and Smart Environments*; Springer: Boston, MA, USA, 2010; pp. 1171–1199.
33. Wolf, P.; Schmidt, A.; Klein, M. SOPRANO-An extensible, open AAL platform for elderly people based on semantical contracts. In Proceedings of the 3rd Workshop on Artificial Intelligence Techniques for Ambient Intelligence (AITAmI'08), 18th European Conference on Artificial Intelligence (ECAI 08), Patras, Greece, 21–25 July 2008.
34. Wolf, P.; Schmidt, A.; Otte, J.P.; Klein, M.; Rollwage, S.; König-Ries, B.; Gabdulkhakova, A. openAAL-the open source middleware for ambient-assisted living (AAL). In Proceedings of the AALIANCE Conference, Malaga, Spain, 11–12 March 2010; pp. 1–5.
35. Adlassnig, K.P.; Blobel, B.; Mantas, J. (Eds.) *Medical Informatics in a United and Healthy Europe: Proceedings of MIE 2009*; IOS Press: Amsterdam, The Netherlands, 2009.
36. Haq, A.N.; Kannan, G. Fuzzy analytical hierarchy process for evaluating and selecting a vendor in a supply chain model. *Int. J. Adv. Manuf. Technol.* **2005**, *29*, 826–835. [CrossRef]
37. Huang, C.C.; Chu, P.Y.; Chiang, Y.H. A fuzzy AHP application in government-sponsored R&D project selection. *Omega* **2008**, *36*, 1038–1052.

Article

Dynamical Simulation of Effective Stem Cell Transplantation for Modulation of Microglia Responses in Stroke Treatment

Awatif Jahman Alqarni [1,2], Azmin Sham Rambely [2,*] and Ishak Hashim [2]

[1] Department of Mathematics, College of Sciences and Arts in Blqarn, University of Bisha, P.O. Box 551, Bisha 61922, Saudi Arabia; aaljhman@ub.edu.sa
[2] Department of Mathematical Sciences, Faculty of Science and Technology, Universiti Kebangsaan Malaysia, UKM Bangi, Selangor 43600, Malaysia; ishak_h@ukm.edu.my
* Correspondence: asr@ukm.edu.my; Tel.: +60-389213244

Abstract: Stem cell transplantation therapy may inhibit inflammation during stroke and increase the presence of healthy cells in the brain. The novelty of this work, is to introduce a new mathematical model of stem cells transplanted to treat stroke. This manuscript studies the stability of the mathematical model by using the current biological information on stem cell therapy as a possible treatment for inflammation from microglia during stroke. The model is proposed to represent the dynamics of various immune brain cells (resting microglia, pro-inflammation microglia, and anti-inflammation microglia), brain tissue damage and stem cells transplanted. This model is based on a set of five ordinary differential equations and explores the beneficial effects of stem cells transplanted at early stages of inflammation during stroke. The Runge–Kutta method is used to discuss the model analytically and solve it numerically. The results of our simulations are qualitatively consistent with those observed in experiments in vivo, suggesting that the transplanted stem cells could contribute to the increase in the rate of ant-inflammatory microglia and decrease the damage from pro-inflammatory microglia. It is found from the analysis and simulation results that stem cell transplantation can help stroke patients by modulation of the immune response during a stroke and decrease the damage on the brain. In conclusion, this approach may increase the contributions of stem cells transplanted during inflammation therapy in stroke and help to study various therapeutic strategies for stem cells to reduce stroke damage at the early stages.

Keywords: cell transplantation; cytokines; ischemic stroke; numerical simulation; runge-kutta method; stability analysis

Citation: Alqarni, A.J.; Rambely, A.S.; Hashim, I. Dynamical Simulation of Effective Stem Cell Transplantation for Modulation of Microglia Responses in Stroke Treatment. *Symmetry* **2021**, *13*, 404. https://doi.org/10.3390/sym13030404

Academic Editor: Sergei D. Odintsov

Received: 8 February 2021
Accepted: 25 February 2021
Published: 2 March 2021

1. Introduction

The numbers of stroke-related deaths are increasing, and globally, stroke is now one of the topmost causes of disability and death [1–3]. During an ischemic stroke, the process by which neurons and glial cells die is known as apoptosis or necrosis [4,5]. Resident microglia are triggered by these dead cells and cause the death of other cells based on environmental toxic substances [5–8]. The components of the brain that prevent the invasion of several diseases are the microglial cells, and these cells can also help prevent stroke [7,9]. Dead cells are phagocytized by activated microglia, which also produce toxic cytokines that impact healthy cells [4,9]. The pathophysiology of ischemic stroke is evident through the inflammatory response [5,10]. Directly after arterial occlusion, inflammation begins in the vasculature and then continues throughout the brain, and systemically throughout the disease stages [10,11]. The body generates tightly regulated immune responses that confer detrimental as well as beneficial properties after stroke [12]. Some of the effects of microglia activation include inhibition of brain repair and/or considerable brain damage, including neurogenesis [11,12]. Due to variability effects on inflammatory processes, the immune response following a stroke serves as a strong determinant of brain restoration or increase the damage in brain [9,11,13]. Thus, inducing stroke recovery-directed modulation of the immune response can offer a potential therapeutic approach [11].

Recently, neurorestorative stem cell-based treatment has been the focus of many stroke studies [1,7]. The extraordinary sophistication of the pathophysiology of ischemic strokes reveals pleiotropic effects on neural stem cells (NSCs), which are potentially therapeutic for both the early (subacute) and chronic phases of stroke [4,7,14]. In addition, after an ischemic stroke for newborn neurons, various obstacles are produced by the generated pathological condition, which makes the use of endogenous repair mechanisms challenging [7,10,15]. Newly formed neurons tend to die, with just $\sim 0.2\%$ survival rate for the remaining cells, while others live up to 5 weeks after ischemia [10].

Microglia can be polarized toward the anti-inflammatory and angiogenic phenotype (M_a) with stem cell transplantation [11]. In one study of ischemic rats, the dependent suppression of inflammation emerged as a classical factor secreted by M_a microglia after the demonstration of intracerebral NSC transplantation and was associated with enhanced angiogenesis and functional recovery in terms of microglia polarization [11]. Currently, it is believed that pro-inflammatory microglia (M_p) can exacerbate brain injury, while anti-inflammatory M_a microglia are neuroprotective [4,12,16,17]. Thus, these dual attributes render them appropriate for use in enhancing post-stroke brain recovery, which can be achieved by shifting their balance from the detrimental M_p to the beneficial M_a phenotype [18]. Evidence also shows that the polarization status of microglia can be altered by stem cells (SCs) and, as such, the observed beneficial actions of SCs can be attributed to skewing microglia toward a neuroprotective and neuroregenerative phenotype [11,14]. Thus, it becomes imperative to elucidate the mechanisms of how SCs respond to tissue damage to understand the crosstalk between inflammation and SCs [10,11]. Additionally, a possible intervention point for regenerative therapies can be met through tweaking the effects of inflammation on stem cell behavior [19].

Studies have shown that stroke SCs can offer a viable solution in the future [11,14,15] and correspondingly, scholars have proposed several models of inflammatory processes [5,20,21]. For example, Di Russo et al. [5] modeled the dynamics of inflammation from a stroke: using (i) necrosis and apoptosis, (ii) the activation and inactivation of resident microglia, and (iii) the ratio of neutrophils and macrophages in the tissue, the dynamics of dead cell densities were investigated. Many scholars have also suggested quantitative methods, such as the use of statistical and computational methods, to study adult neurogenesis [22–24]. Along the same lines, systems of a hierarchical cell-constructing model were studied by Nakata et al. [25], where the structure system was controlled by the adult cells. Ziebell et al. [23] introduced a mathematical model that portrayed the various stages of the adult hippocampus and the evolving dynamics of SCs.

The basic framework for our study is a mathematical model of the interplay between microglia and endogenous NSCs during a stroke. Alqarni et al. [4] evaluated potential mechanisms to regulate and stabilize the treatment of microglia inflammation associated with an exogenous stem cell transplantation stroke. Alharbi and Rambely used ordinary differential equations in the formulation of mathematical models to describe the effect of vitamins on strengthening the immune system and its function in delaying the development of tumour cells [26–28]. In addition, Ordinary differential equations (ODEs) have been used to explain disease behavior over time, which has improved therapeutic strategies [4,28,29].

We aim to enhance understanding of the symmetry and antisymmetry of the relationship between the functions of pro-inflammatory and anti-inflammatory microglia with the effect of transplanted stem cells on the immunomodulation of microglia functions during a stroke. We modified a mathematical model for the therapy of the inflammation process in stroke using ODEs. In this paper, we study the roles of stem cell transplantation dynamics via inhibition of inflammation from microglia and stimulate the beneficial function of microglia during a stroke. We investigate the possible mechanisms involved in the dynamics of the transplanted stem cell functions through immune activities.

This paper is organized as follows: In Section 2, a mathematical model called SCs–damage–resting microglia–pro-inflammation microglia– anti-inflammation microglia (SDRPA) is presented. In Section 3, we study the model's equilibrium points. In Section 4, we investigate the model's stability. In Section 5, numerical experiments are discussed. Finally, the study's conclusions are presented in Section 6.

2. Mathematical Representation of the SDRPA Model

In this section, we give a dynamic model for the explanation of the use of transplanted SCs to treat the inflammation produced by microglia during a stroke. The suggested model of therapy stem cell dynamics transplanted in onset stroke shows the interplay between the transplanted SCs and microglia in stroke. The mathematical model is a structure of five differential equations, which are analyzed to find the equilibrium points and to study their stability. Several previous studies have shown the possible therapeutic benefits of transplanted SCs, where the transplantation of pluripotent stem cells in stroke patients has many functions. One of the potential therapeutic of this type of treatment is the immunomodulation of microglia functions during a stroke [11,30].

There are three options for differentiating SCs [31] : (1) symmetrical self-renewal with the possibility σ_S of two SCs, (2) asymmetric self-renewal with the probability of σ_A, where one of the cells denotes the daughter residue a stem cell, whereas another cell does not make this discrimination, and (3) symmetric involvement differentiation with probability of σ_D, where a stem cell has the ability to divide to be two involvement cells. Here we suppose that $\sigma_S + \sigma_A + \sigma_D = 1$ [31]. In this manuscript, we assumed that SCs S reproduce at the rate σ and die at the rate γ_s. Thus, transplanted SCs have considerable influence on the neuroinflammation caused by a stroke because they secrete relevant cytokines to support the anti-inflammation of microglial activation to transform the M_a to M_p phenotype [32]. Therefore, the terms α_4, α_5 and α_6 were designated to describe the direct interactions between SCs and the microglia M_a and M_p. The following equation describes transplanted SCs' behavior during a stroke:

$$\frac{dS}{dt} = [\sigma - \alpha_5 M_a - \alpha_6 M_p - \gamma_s]S.$$

where σ indicates the reproduction rate of SCs, α_5 signifies the stimulating and supporting rate of SCs for M_a, while α_6 indicates the rate of elimination of SCs due to the M_p and γ_s indicates the rate of death of SCs.

During an ischemic stroke, microglia are reactivated and polarized to either a classical type, M_p, by pro-inflammatory cytokines that cause an immune response and lead to secondary damage in the brain; or to a substitutional type, M_a, an anti-inflammatory state caused through anti-inflammatory cytokines, that reduces inflammation and boosts cell repair [33–36]. For mathematical modeling purposes, we supposed that resting immune cells of microglia M_r were produced at a constant rate of α_0, indicating the source rate of the resting microglia cells, which are dying at a constant ratio of γ_0.

The given differential equation describes resting microglia cells' behavior:

$$\frac{dM_r}{dt} = \alpha_0 - (\alpha_1 + \alpha_2 + \gamma_0)M_r.$$

Microglia that are activated by cytokines are caused through dead cells: the M_p phenotype is activated through signals of pro-inflammation cytokine α_1 and the M_a phenotype is activated by signals of anti-inflammation cytokine α_2. Moreover, a shift from M_p to M_a occurs at the rate of parameter α_3, where α_3 is the beginning of transference from pro-inflammatory microglia to anti-inflammatory microglia. The modulation of the immune response is considered an important function of a possible therapeutic approach to improve brain recovery post-stroke [11,19]. Exogenous stem cell transplantation has been demonstrated to modulate the inflammatory immune microenvironment across the ischemic regions of the brain by modulating the functions of the immune cells during the stroke [11,19,33]. Direct transplant of SCs into the brain after ischemia decreased many inflammatory and immune responses and switched the balance from a pro-inflammatory to anti-inflammatory response of microglia [37]. We present these functions by the rate of parameters α_4 and α_5. The following equations representing microglia M_a and M_p thus take the form:

$$\begin{aligned}\frac{dM_p}{dt} &= \alpha_1 M_r - (\alpha_4 S + \delta_1 D + \alpha_3 + \gamma_1) M_p, \\ \frac{dM_a}{dt} &= \alpha_2 M_r + (\alpha_5 S - \gamma_2) M_a + (\alpha_3 + \alpha_4 S) M_p.\end{aligned}$$

where, δ_1 signifies the rate of damage from microglia related to the production of cytokines, α_4 denotes the rate of amendment of SCs for the function of M_p, and γ_1 and γ_2 are the natural death rate of M_p and M_a, respectively.

Ischemic stroke injury is regarded as a major factor contributing to tissue damage. We assume that the SDRPA model shows that activated microglial cells play complex functions displaying both harmful and beneficial effects, which could include from elimination of cell debris by phagocytosis process and the release of inflammatory cytokines that lead to tissue destruction and increase cell death beyond the primary ischemic region [12,34]. Therefore, the dynamics of dead brain cells after stroke can be explained via the differential equation:

$$\frac{dD}{dt} = [(\delta_2 - r_1) M_p - r_2 M_a] D.$$

where δ_2 indicates the death rate of brain cells due to M_p; and r_1 and r_2 refers to the elimination rates of the damaged cells by M_p and M_a, respectively. Thus, the SDRPA model can be written in the following form:

$$\frac{dS}{dt} = [\sigma - \alpha_5 M_a - \alpha_6 M_p - \gamma_s] S, \tag{1}$$

$$\frac{dD}{dt} = [(\delta_2 - r_1) M_p - r_2 M_a] D, \tag{2}$$

$$\frac{dM_r}{dt} = \alpha_0 - (\alpha_1 + \alpha_2 + \gamma_0) M_r, \tag{3}$$

$$\frac{dM_p}{dt} = \alpha_1 M_r - (\alpha_4 S + \delta_1 D + \alpha_3 + \gamma_1) M_p, \tag{4}$$

$$\frac{dM_a}{dt} = \alpha_2 M_r + (\alpha_5 S - \gamma_2) M_a + (\alpha_3 + \alpha_4 S) M_p. \tag{5}$$

The initial conditions in this model are: $S(0) = S^*$, $D(0) = D^*$, $M_r(0) = M_r^*$, $M_p(0) = M_p^*$, and $M_a(0) = M_a^*$, $0 \leq t < \infty$ where $S(t)$, $D(t)$, $M_r(t)$, $M_p(t)$, and $M_a(t)$ represent the stem cell concentration, dead cells, resting microglia, pro-inflammation microglia, and the anti-inflammation microglia, respectively.

The proposed dynamic model of SDRPA demonstrated by (1)–(5), represents the population behavior of the transplanted SCs, immune cells, and damaged cells in a stroke. Thus, the variables $S(t)$, $D(t)$, $M_r(t)$, $M_p(t)$, and $M_a(t)$ and all parameters are non-negative real, real, and equal to or less than one. The invariant area of SDRPA becomes:

$$\Psi = (S, D, M_r, M_p, M_a) \in \Re^5_+ \tag{6}$$

3. The SDRPA Model's Equilibrium Points

We determine the fixed points of the SDRPA system (1)–(5) from the following:

- $\frac{dS}{dt} = 0 \Leftrightarrow$

$$[\sigma - \alpha_5 M_a - \alpha_6 M_p - \gamma_s] S = 0, \tag{7}$$

- $\frac{dD}{dt} = 0 \Leftrightarrow$

$$[(\delta_2 - r_1)M_p - r_2 M_a]D = 0, \tag{8}$$

- $\frac{dM_r}{dt} = 0 \Leftrightarrow$

$$\alpha_0 - (\alpha_1 + \alpha_2 + \gamma_0)M_r = 0, \tag{9}$$

- $\frac{dM_p}{dt} = 0 \Leftrightarrow$

$$\alpha_1 M_r - (\alpha_4 S + \delta_1 D + \alpha_3 + \gamma_1)M_p = 0, \tag{10}$$

- $\frac{dM_a}{dt} = 0 \Leftrightarrow$

$$\alpha_2 M_r + (\alpha_5 S - \gamma_2)M_a + (\alpha_3 + \alpha_4 S)M_p = 0. \tag{11}$$

The equilibrium points of SDRPA (1)–(5) calculate by solving Equations (7)–(11), we obtain three positive real points of equilibrium by using MATHEMATICA, while the other solutions will be negative. Thus, the first positive equilibrium point is given by:

$$Q_1(S, D, M_r, M_p, M_a) = \left(0, 0, \frac{\alpha_0}{z}, \frac{\alpha_0\alpha_1}{z(\alpha_3 + \gamma_1)}, \frac{\alpha_0(\alpha_1\alpha_3 + \alpha_2(\alpha_3 + \gamma_1))}{z(\alpha_3 + \gamma_1)\gamma_2}\right).$$

where

$$z = \alpha_1 + \alpha_2 + \gamma_0. \tag{12}$$

Next, we represent the second positive equilibrium point as follows:

$$Q_2(S, D, M_r, M_p, M_a) = \left(0, \frac{q}{r_2\,\alpha_2\,\delta_1}, \frac{\alpha_0}{z}, -\frac{r_2\alpha_0\alpha_2}{z\,p}, \alpha_0\frac{\alpha_2(r_1 - \delta_2)}{z\,p}\right),$$

where

$$\begin{aligned} p &= (r_2\alpha_3 + \gamma_2(r_1 - \delta_2)) < 0, \\ q &= -r_2(\alpha_1\alpha_3 + \alpha_2(\alpha_3 + \gamma_1)) + \alpha_1\gamma_2(-r_1 + \delta_2). \end{aligned}$$

The third positive equilibrium point is obtained as follows:

$$Q_3(S, D, M_r, M_p, M_a) = \left(\frac{\beta_3}{\beta_1}, \beta_2, \frac{\alpha_0}{z}, r_2\beta_0, (\delta_2 - r_1)\beta_0\right),$$

where

$$\begin{aligned} \beta_0 &= \frac{\omega}{\alpha_5(-r_1 + \delta_2) + r_2\alpha_6} > 0, \beta_1 = (r_2\alpha_4 + \alpha_5(-r_1 + \delta_2)) > 0, \\ \beta_2 &= \frac{1}{r_2 z\delta_1\beta_1\omega}[(r_1^2 + \delta_2^2)\alpha_0\alpha_1\alpha_5^2 + r_2^2\alpha_4(\alpha_0(\alpha_1 + \alpha_2)\alpha_6 - z\gamma_1\omega) \\ &\quad + r_2(-r1 + \delta_2)(\alpha_0\alpha_5(\alpha_2\alpha_4 + \alpha_1(\alpha_4 + \alpha_6))) - r_1(2\alpha_0\alpha_1\alpha_5^2\delta_2) \\ &\quad -(-r_1 + \delta_2)r_2 z(\alpha_5(\alpha_3 + \gamma_1) + \alpha_4\gamma_2)\omega] < 0, \\ \beta_3 &= -p - \frac{\alpha_0\alpha_2(r_2\alpha_6 + \alpha_5(\delta_2 - r_1))}{z\omega} > 0, \omega = -\gamma_s + \sigma > 0. \end{aligned}$$

Proposition 1. *We assume that the equilibrium points for the SDRPA system, $S; D; M_r; M_p; M_a > 0$ are satisfied under the conditions:*

- $r_1 < \delta_2$
- $\gamma_s < \sigma$

- $\frac{\alpha_0\alpha_2(r_2\alpha_6+\alpha_5(\delta_2-r_1))}{z\omega} < \gamma_2(-r_1+\delta_2) - r_2\alpha_3$
- $r_2\alpha_3 < \gamma_2(\delta_2 - r_1))$
- $r_2(\alpha_1\alpha_3 + \alpha_2(\alpha_3 + \gamma_1)) < \gamma_2\alpha_1(\delta_2 - r_1)$

Non-negative real, steady states will then, and only then, exist.

We can study the stability of equilibrium points if they continue to exist constantly over time or constantly change in equilibrium in one direction. We defined three steady states as follows:

Definition 1. *We define the resting microglia activation in a normal brain as the absence of a high activation for these cells when stroke occurs into pro-inflammation microglia and anti-inflammation microglia, the steady-state* $M_r; M_p; M_a > 0$ *and* $D, S = 0$ *implies the functions of the microglia in the onset of a stroke is normal and the microglia do not cause any damage in the brain.*

Definition 2. *We define the starting of the damage by the increased rate of pro-inflammation cytokines where the existence of high activation of microglia will cause damage in brain, the steady-states* $D; M_r; M_p; M_a > 0$, *and* $S = 0$ *imply the damage of the brain cells from the pro-inflammation of microglia by the damaged brain cells from inflammation by microglia due to a stroke without stem cells transplantation.*

Definition 3. *We define the functions of transplanted stem cells on modulation of microglia responses in the brain after stroke, the steady-state of the forms* $S; D; M_r; M_p; M_a > 0$ *is defined by the role of stem cell transplantation in inflammation.*

4. The Equilibrium Points' Stability of the SDRPA Model

A stability study of the SDRPA model for the equilibrium points is performed. By applying the Hartman–Grobman theorem concept [38], the system's 5×5 Jacobian matrix for the eigenvalues associated with transplanted SCs equilibrium in the brain after a stroke (1)–(5) is given by:

$$J[\tau] = \begin{bmatrix} F_S[\varphi] & F_D[\varphi] & F_{M_r}[\varphi] & F_{M_p}[\varphi] & F_{M_a}[\varphi] \\ G_S[\varphi] & G_D[\varphi] & G_{M_r}[\varphi] & G_{M_p}[\varphi] & G_{M_a}[\varphi] \\ H_S[\varphi] & H_D[\varphi] & H_{M_r}[\varphi] & H_{M_p}[\varphi] & H_{M_a}[\varphi] \\ K_S[\varphi] & K_D[\varphi] & K_{M_r}[\varphi] & K_{M_p}[\varphi] & K_{M_a}[\varphi] \\ L_S[\varphi] & L_D[\varphi] & L_{M_r}[\varphi] & L_{M_p}[\varphi] & L_{M_a}[\varphi] \end{bmatrix}.$$

where $\varphi = [S, D, M_r, M_p, M_a], F[\varphi] = \frac{dS}{dt}$, $G[\varphi] = \frac{dD}{dt}, H[\varphi] = \frac{dM_r}{dt}, K[\varphi] = \frac{dM_p}{dt}$, and $L[\varphi] = \frac{dM_a}{dt}$.

Theorem 1. *Given that the function* $g : \Psi \to \Re^5_+$ *where* Ψ *is a domain in* $\Re^5_+$, *and assuming that* $Q_1 \in \Psi$ *is an equilibrium point, where one eigenvalue of the Jacobian matrix has a non-negative real part at minimum. Therefore,* Q_1 *indicates an unstable equilibrium point of* g.

Proof. The Jacobian matrix J calculated at the first equilibrium point Q_1 yields: $J[Q_1] = \begin{bmatrix} a_{11} & 0 & 0 & 0 & 0 \\ 0 & a_{22} & 0 & 0 & 0 \\ 0 & 0 & -z & 0 & 0 \\ -\frac{\alpha_0\alpha_1\alpha_4}{z(\alpha_3+\gamma_1)} & -\frac{\alpha_0\alpha_1\delta_1}{z(\alpha_3+\gamma_1)} & \alpha_1 & -\alpha_3-\gamma_1 & 0 \\ a_{15} & 0 & \alpha_2 & \alpha_3 & -\gamma_2 \end{bmatrix}$ where

$$\begin{aligned} a_{11} &= -\frac{\alpha_0(\alpha_2\alpha_5(\alpha_3+\gamma_1)+\alpha_1(\alpha_3\alpha_5+\alpha_6\gamma_2))}{z(\alpha_3+\gamma_1)\gamma_2} - \gamma_s + \sigma, \\ a_{22} &= -\frac{r_2\alpha_0(\alpha_1\alpha_3+\alpha_2(\alpha_3+\gamma_1))}{z(\alpha_3+\ \gamma_1)\gamma_2} + \frac{\alpha_0\alpha_1(-r_1+\delta_2)}{z(\alpha_3+\gamma_1)}, \\ a_{15} &= \frac{\alpha_0\alpha_1\alpha_4}{z(\alpha_3+\gamma_1)} + \frac{\alpha_0\alpha_5(\alpha_1\alpha_3+\alpha_2(\alpha_3+\gamma_1))}{z(\alpha_3+\gamma_1)\gamma_2}, \\ z &= (\alpha_1+\alpha_2+\gamma_0). \end{aligned}$$

We determine the eigenvalues of the matrix $J[Q_1]$ as follows:

$$\begin{aligned} \lambda_1 &= -z < 0, \\ \lambda_2 &= -\frac{\alpha_0(r_2(\alpha_1+\alpha_2)\alpha_3 + r_2\alpha_2\gamma_1 + \alpha_1\gamma_2(r_1-\delta_2))}{z(\alpha_3+\ \gamma_1)\gamma_2} > 0, \\ \lambda_3 &= -\frac{\alpha_0(\alpha_2\alpha_5(\alpha_3+\gamma_1)+\alpha_1(\alpha_3\alpha_5+\alpha_6\gamma_2))}{z(\alpha_3+\gamma_1)\gamma_2} - \gamma_s + \sigma < 0, \\ \lambda_4 &= -\gamma_2 < 0, \lambda_5 = -(\alpha_3+\gamma_1) < 0. \end{aligned}$$

From Proposition 1, the equilibrium points, it is obvious that $\lambda_2 > 0$. Thus, $J(Q_1)$ has at least one positive root, which denotes that the second equilibrium point Q_1 is unstable. □

Theorem 2. *Given the function $g : \Psi \to \Re^5_+$, where Ψ is a domain in $\Re^5_+$, and assuming that $Q_2 \in \Psi$ is an equilibrium point, then one eigenvalue of the Jacobian matrix has a non-negative real part at minimum. Therefore, Q_2 indicates an unstable equilibrium point of g.*

Proof. The Jacobian matrix J calculated at the second equilibrium point Q_2 yields: $J[Q_2] = \begin{bmatrix} b_{11} & 0 & 0 & 0 & 0 \\ 0 & 0 & 0 & b_{24} & -\frac{q}{\alpha_2\delta_1} \\ 0 & 0 & -z & 0 & 0 \\ v\alpha_4 & v\delta_1 & \alpha_1 & \frac{\alpha_1 p}{r_2\alpha_2} & 0 \\ -\frac{\alpha_0\alpha_2\beta_1}{Zp} & 0 & \alpha_2 & \alpha_3 & -\gamma_2 \end{bmatrix}$ where

$$b_{11} = \omega + \frac{\alpha_0\alpha_2(-r_1\alpha_5+r_2\alpha_6+\alpha_5\delta_2)}{zp}, v = \frac{r_2\alpha_0\alpha_2}{zp}, b_{24} = \frac{(-r_1+\delta_2)q}{r_2\alpha_2\delta_1}.$$

The corresponding characteristic equation of the Jacobian $J[Q_2]$, is specified by:

$$(b_{11}-\lambda)(-z-\lambda)(\kappa_0+\kappa_1\lambda+\kappa_2\lambda^2+\lambda^3) = 0.$$

where

$$\kappa_2 = -\frac{p\,\alpha_1}{r_2\,\alpha_2} + \gamma_2, \kappa_1 = \frac{-p\,\alpha_1\,\gamma_2 + q\,v\,(r_1-\delta_2)}{r_2\,\alpha_2}, \kappa_0 = \frac{q\,v\,p}{r_2\,\alpha_2}.$$

Thus, we can obtain the first two eigenvalues:

$$\lambda_1 = \omega + \frac{\alpha_0\alpha_2(-r_1\alpha_5+r_2\alpha_6+\alpha_5\delta_2)}{zp} > 0, \lambda_2 = -z < 0.$$

From Proposition 1, the equilibrium points, it is obvious that $\lambda_1 > 0$. Thus, $J(Q_2)$ has at least one positive root, which denotes that the second equilibrium point Q_2 is unstable. □

Theorem 3. *Given the function $g : \Psi \to \Re^5_+$, where Ψ denotes a domain in $\Re^5_+$, and assuming that $Q_3 \in \Psi$ indicates an equilibrium point, all the Jacobian matrix's eigenvalues have negative real parts at the equilibrium point Q_3. Therefore, Q_3 is assumed as a stable equilibrium point of g.*

Proof. The Jacobian matrix J calculation at the third equilibrium point Q_3 yields: $J[Q_3] = \begin{bmatrix} 0 & 0 & 0 & c_{14} & c_{15} \\ 0 & 0 & 0 & c_{24} & c_{25} \\ 0 & 0 & -z & 0 & 0 \\ c_{41} & c_{42} & \alpha_1 & c_{44} & 0 \\ c_{51} & 0 & \alpha_2 & c_{54} & c_{55} \end{bmatrix}$ where

$$\begin{aligned} c_{14} &= -\frac{\beta_3}{\beta_1}\alpha_6 < 0, c_{24} = \beta_2(-r_1 + \delta_2) > 0 \\ c_{15} &= -\frac{\beta_3}{\beta_1}\alpha_5 < 0, c_{25} = -\beta_2 r_2 < 0, c_{41} = -\beta_0 r_2 \alpha_4 < 0, \\ c_{42} &= -\beta_0 r_2 \delta_1 < 0, c_{44} = -\alpha_3 - \frac{\beta_3}{\beta_1}\alpha_4 - \gamma_1 - \beta_2\delta_1 < 0, \\ c_{51} &= \beta_0\beta_1 > 0, c_{54} = \alpha_3 + \frac{\beta_3}{\beta_1}\alpha_4 > 0, \\ c_{55} &= \frac{\beta_3}{\beta_1}\alpha_5 - \gamma_2 < 0. \end{aligned}$$

The characteristic equation can be written as:

$$(z + \lambda)(\lambda^4 + \eta_3\lambda^3 + \eta_2\lambda^2 + \eta_1\lambda + \eta_0) = 0, \tag{13}$$

Here,

$$\begin{aligned} \eta_0 &= c_{42}c_{51}M_0 > 0, \eta_1 = c_{44}M_6 + c_{41}M_4 - c_{42}M_5 > 0, \\ \eta_2 &= M_1 + M_2 > 0, \eta_3 = -M_3 > 0. \end{aligned}$$

where,

$$\begin{aligned} M_0 &= c_{15}c_{24} - c_{14}c_{25} < 0, M_1 = c_{14}c_{41} - c_{24}c_{42} > 0, \\ M_2 &= -c_{15}c_{51} + c_{44}c_{55} > 0, M_3 = c_{44} + c_{55} < 0, \\ M_4 &= c_{15}c_{54} - c_{14}c_{55} < 0, M_5 = c_{25}c_{54} - c_{24}c_{55} > 0, \\ M_6 &= c_{15}c_{51} < 0. \end{aligned}$$

Thus, we can determine the first eigenvalue:

$$\lambda_1 = -z < 0.$$

Thus, we can use the Routh–Hurwitz theorem for $\lambda^4 + \eta_3\lambda^3 + \eta_2\lambda^2 + \eta_1\lambda + \eta_0 = 0$ [39], giving

$$\begin{vmatrix} \lambda^4 & 1 & \eta_2 & \eta_0 \\ \lambda^3 & \eta_3 & \eta_1 & 0 \\ \lambda^2 & \xi_1 & \eta_0 & 0 \\ \lambda^1 & \xi_2 & 0 & 0 \\ \lambda^0 & \eta_0 & 0 & 0 \end{vmatrix}.$$

Thus, the essential and adequate conditions of all roots contain negative real parts η_i $(i = 1;3;4) > 0$ and $\Delta > 0$. Then,

$$\begin{aligned}\Delta &= c_{42}c_{51}M_0(-c_{42}c_{51}M_0M_3^2 - (M_1 + M_2)M_3(c_{41}M_4 - c_{42}M_5 \\ &+ c_{44}M_6) - (c_{41}M_4 - c_{42}M_5 + c_{44}M_6)^2) > 0\end{aligned}$$

From Proposition 1,

$$\xi_1 = \frac{\eta_3\eta_2 - \eta_1}{\eta_3} > 0, \xi_2 = \frac{\xi_1\eta_1 - \eta_0\eta_3}{\xi_1} > 0$$

where

$$\begin{aligned}\xi_1 &= \frac{(M_1 + M_2)M_3 + c_{41}M_4 - c_{42}M_5 + c_{44}M_6}{M_3} > 0 \\ \xi_2 &= c_{41}M_4 - c_{42}M_5 + c_{44}M_6 + \frac{c_{42}c_{51}M_0M_3^2}{(M_1 + M_2)M_3 + c_{41}M_4 - c_{42}M_5 + c_{44}M_6} > 0\end{aligned}$$

Equation (13) has no roots with positive real parts, and only one of its eigenvalues is negative in view of the positive signs of all the coefficients in the first column. The equilibrium point Q_3 is therefore stable. □

Remark 1. *The impact of using SCs transplant on the functions of microglia in stroke onset, represented by the dynamic SDRPA model, can be summarized as follows:*

- *Theorem 1 indicates that the damage, D, can penetrate the SDRPA model, if $\lambda_2 > 0$.*
- *Theorem 2 indicates that the damage, $D > 0$, penetrated the brain.*
- *Theorem 3 indicates that stem cell transplantation, $S > 0$, modulates the inflammatory environment in a stroke, $D > 0$.*
- *The SDRPA model is considered stable when the immunomodulation from transplanted stem cells can be one of the mechanisms of post-stroke recovery.*

5. Numerical Results and Analysis

This section discusses the utilization of computational models to assess which ones affect the model's behavior and thus explore the numerical solutions of the system ((1)–(5)).

5.1. Determination of Parameters

The SDRPA model involves 18 parameters, including five parameters for the initial conditions of each compartment. Herein, the researchers list several parameters which can be assessed using the experimental studies. Furthermore, for solving the system of ordinary differential equations ((1)–(5)), other parameter values were obtained by simulations through the Software MATHEMATICA with the command NDSolve. The purpose was to assess the SCs' ability to control the functions of immune cells during a stroke. From the simulation of SDRPA model, it can be concluded that there are three parameters that directly affect the behavior of microglia and SCs after transplant, parameter α_4, which represents the rate of amendment of SCs for the function of M_p, parameter α_5, which shows the rate of stimulating and supporting SCs of anti-inflammatory immune cells and, the suppression rate of SCs as a result of damage by pro-inflammatory microglia denoted by the parameter α_6, and the simulation results showed that the effectiveness of stem cell transplantation to modulate pro-inflammation microglia function where the highest rate of SCs will switch the balance from a pro-inflammatory to anti-inflammatory response of microglia as follows: the rate involving SCs was simulated by $\alpha_4 = 50\%, \alpha_5 = 35\%$ and $\alpha_6 = 14\%$. Furthermore, it was obvious that the modulation of microglia responses occurred if the following conditions were satisfied: The rate of stem cell amendment for the function of $M_p >$; and the suppression rate of SCs by M_p, $\alpha_4 > \alpha_6$. Table 1 displays the reference series of values of the parameters. From Figure 1, we can observe various behaviors depending on the transplant process. In addition, damage continually decreased, while pro-inflammation microglia had a little increase, asides from shifting to a steep curve after approximately 20 h. Furthermore, that anti-inflammation microglia had a high level after stroke onset and transplantation SCs, aside from shifting to a steep and a stable curve at approximately after 20 h to the end of simulation of the recovery stage. It can be seen that after SCs transplanted in the brain the anti-inflammation microglia had a higher level than the pro-inflammation microglia comparing with numerical results evaluating microglia effect on the brain dynamics without SCs transplantation during 72 h of strok as presented in [4]. Additionally, for the concentration of transplanted SCs, there was initial growth and subsequently a reduction over time due to the effects of the treatment on the elimination rate of pro-inflammation microglia. The system was solved numerically, and time concatenation of the solutions were plotted on the system ((1)–(5)) for the parameters to obtain the dynamics of the system. For this, the fourth-order Runge–Kutta method (RK4) was used in all simulations to obtain extra stable and approximate solutions. The time 72 h was chosen for 10^{-4} as the step size for carrying out the simulations of the model [40,41]. Figures 2 and 3 depict the precision of the suggested numerical method presented through the residual error.

Table 1. Parameters values for transplanted stem cells–damaged brain cells–resting microglia–pro-inflammation microglia–anti-inflammation microglia (SDRPA) model.

Parameters	Values	Descriptions	Sources
S^*	1	SCs initial concentration	[31]
D^*	0.4	damage initial concentration	[4]
M_r^*	1	resting microglia initial concentration	[4]
M_p^*	0.1415	pro-inflammation initial concentration	[4]
M_a^*	0.02	anti-inflammation initial concentration	[4]
σ	0.69	the reproduction rate of stem cells	[42]
α_0	0.38	the resting microglia source	[4]
α_1	0.12	activation rate of M_r into M_p	[4]
α_2	0.017	activation rate of M_r into M_a	[4]
α_3	0.11	the rate transference from M_p to M_a	[4]
δ_1	0.2854	the cytotoxic effects due to M_p	[5]
δ_2	0.1	the death rate of brain cells due to M_p	[5]
γ_s	0.1	the natural death rate of S	[42]
γ_0	0.003	the natural death rate of M_r	[4]
γ_1	0.06	the natural death rate of M_p	[4]
γ_2	0.05	the natural death rate of M_a	[4]
r_1	0.05	the decay rate of concentration of the D by M_p	[4]
r_2	0.0125	the decay rate of concentration of the D by M_a	[4]

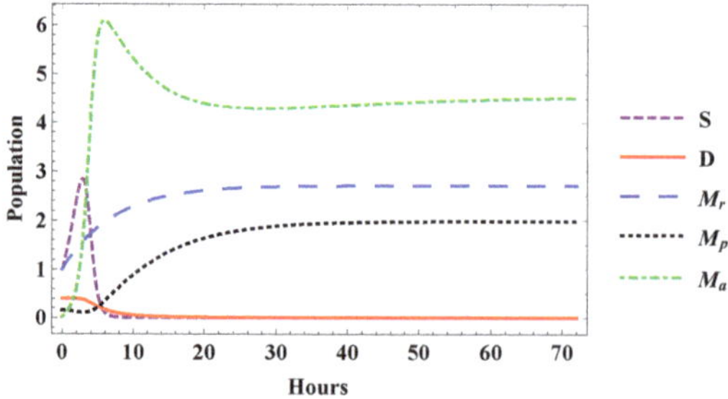

Figure 1. Effectiveness of stem cell transplantation on immune cells' phenotype behavior and dead brain cells within 72 h. Where $\alpha_4 = 0.5314, \alpha_5 = 0.3468$ and $\alpha_6 = 0.1419$.

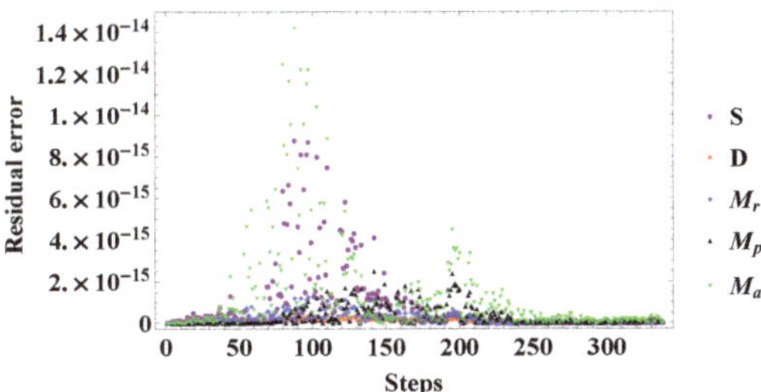

Figure 2. The residual error at several steps for SDRPA model.

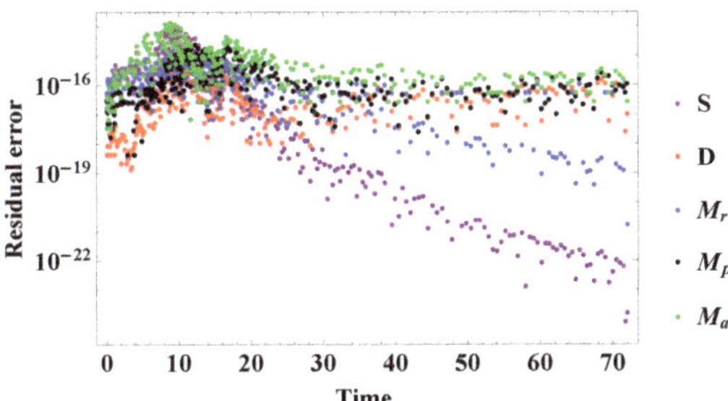

Figure 3. The residual error at time t for SDRPA model.

5.2. Comparison of Experimental Results

Parameter values of resting microglia rates, activation microglia rates, decay rates of concentration of the damage by microglia, reproduction rate of SCs, death rates, and initial conditions, were assumed on the basis of the literature. We assumed that the SCs transplantation influenced transference between the states of microglia activation and could support the anti-inflammation function of microglia, which led to faster recovery. These assumptions were in accordance with the biological understanding of SC functions. In contrast to the findings of our study regarding immune cell functions and the damage during the time of stroke with the SCs effects transplanted with the work in [16,34], we established that the studies agreed in that microglial activation had the contribution of both beneficial and harmful functions in the brain. Furthermore, stem cell transplantation is considered to attenuate ischemia-induced brain injury within hours of transplantation [1,33,40]. The simulation's findings of the model corresponded with the experimental findings for determining the effect of exogenous SCs in the brain during a stroke. We compared the simulation results of the SDRPA model with the numerical results of the effects of dynamic SCs on brain therapy after stroke by endogenous SCs and exogenous SCs. Additionally, we compared the findings in our research to those of [4], and found that stem cell transplantation has the ability to modify the cytokine environment in the brain, especially for early cell transplantation after stroke. For example, transplantation within 1 week caused a reduction in the levels of pro-inflammatory cytokines and growth in the levels of anti-inflammatory cytokines. The results of the SDRPA simulation implied that when neural endogenous SCs failed to grow and assist brain recovery, exogenous SCs were considered the best solution for brain repair processes.

6. Conclusions

Mathematics and computer science fields have worked interactively to better understand biological processes. We modified the model—SDRPA ((1)–(5)) based on the model of Alqarni et al. [4] to study the effectiveness of the role of transplanted exogenous SCs in the brain on the microglia during a stroke. The multifaceted SCs affect the tissue by inhibiting of the pro-inflammation and stimulating the function of the anti-inflammation microglia. We discussed the results of the dynamical system and the effects of SCs transplanted and microglia during the stroke analytically and numerically. The analysis and simulation results of the system show the ability of transplanted SCs to help the brain by reducing the inflammation on the onset of stroke. Following that, the rate of damage reduced after transplantation. An evident linkage between the mathematical and biological mechanisms was observed. From the analytical results, we can deduce that the stability of the SDRPA model illustrated capacity in the exogenous stem cell implantation, which is significant for immunomodulation. In conclusion, our model may assist in conception of the effectiveness of using SCs transplanted in the brain repair processes. In the outlook, we will extend this study to model strategies that improve, stimulate, and generate the NSCs in the early stage, and where that information could contribute to understanding the effects of therapeutic strategies. We hope to conduct more experimental studies to clinically investigate the results of our mathematical model and to show more precise results. In future studies, it could be interesting to incorporate the dynamics of anti-inflammatory and pro-inflammatory cytokines from microglia and cytokines of endogenous NSCs into the SDRPA model to describe the interaction processes of the different cytokine types in ischemic stroke. Furthermore, we will develop this model by studying the effect of SCs in stimulating endogenous neural cells in a stroke and dynamically validate the ability to support the endogenous stem cells by using pharmacological drugs to improve stroke therapy. In summary, symmetry and antisymmetry are fundamental to understand the role of microglia in ischemic stroke pathobiology constituting a major challenge for the development of efficient immunomodulatory therapies by SCs.

Author Contributions: Conceptualization, A.J.A.; funding acquisition, A.S.R.; methodology, A.J.A.; project administration, A.S.R.; supervision, A.S.R. and I.H.; validation, A.S.R.; writing—original draft, A.J.A.; writing—review and editing, I.H. All authors have read and agreed to the published version of the manuscript.

Funding: This research is funded by a grant from Universiti Kebangsaan Malaysia with code GUP-2017-112.

Institutional Review Board Statement: Not applicable.

Informed Consent Statement: Not applicable.

Data Availability Statement: Not applicable.

Acknowledgments: The authors would like to thank the Universiti Kebangsaan Malaysia for the support in this study.

Conflicts of Interest: All authors declare that there is no conflict of interest associated with this study.

References

1. Marei, H.E.; Hasan, A.; Rizzi, R.; Althani, A.; Afifi, N.; Cenciarelli, C.; Shuaib, A. Potential of Stem Cell-Based Therapy for Ischemic Stroke. *Front. Neurol.* **2018**, *9*, 34. [CrossRef]
2. Nordin, N.A.M.; Aljunid, S.M.; Aziz, N.A.; Nur, A.M.; Sulong, S. Direct medical cost of stroke: Findings from a tertiary hospital in malaysia. *Med. J. Malays.* **2012**, *67*, 473–477.
3. Chouw, A.; Triana, R.; Dewi, N.M.; Darmayanti, S.; Rahman, M.N.; Susanto, A.; Putera, B.W.; Sartika, C.R. Ischemic Stroke: New Neuron Recovery Approach with Mesenchymal and Neural Stem Cells. *Mol. Cell. Biomed. Sci.* **2018**, *2*, 48–54. [CrossRef]
4. Alqarni, A.J.; Rambely, A.S.; Hashim, I. Dynamic Modelling of Interactions between Microglia and Endogenous Neural Stem Cells in the Brain during a Stroke. *Mathematics* **2020**, *8*, 132. [CrossRef]
5. Russo, C.D.; Lagaert, J.-B.; Chapuisat, G.; Dronne, M.-A. A mathematical model of inflammation during ischemic stroke. *ESAIM Proc.* **2010**, *30*, 15–33. [CrossRef]
6. Hui-Yin, Y.; Ahmad, N.; Azmi, N.; Makmor-Bakry, M. Curcumin: The molecular mechanisms of action in inflammation and cell death during kainate-induced epileptogenesis. *Indian J. Pharm. Educ. Res.* **2018**, *52*, 32–41. [CrossRef]
7. Vay, S.U.; Flitsch, L.J.; Rabenstein, M.; Rogall, R.; Blaschke, S.; Kleinhaus, J.; Reinert, N.; Bach, A.; Fink, G.R.; Schroeter, M.; et al. The plasticity of primary microglia and their multifaceted effects on endogenous neural stem cells in vitro and in vivo. *J. Neuroinflamm.* **2018**, *15*, 226. [CrossRef]

8. Rajab, N.F.; Musa, S.M.; Munawar, M.A.; Mun, L.L.; Yen, H.K.; Ibrahim, F.W.; Meng, C.K. Anti-neuroinflammatory Effects of Hibiscus sabdariffa Linn. (Roselle) on Lipopolysaccharides-induced Microglia and Neuroblastoma Cells. *J. Sains Kesihat. Malays.* **2016**, *14*, 111–117. [CrossRef]
9. Galloway, D.A.; Phillips, A.E.M.; Owen, D.R.J.; Moore, C.S. Phagocytosis in the brain: Homeostasis and disease. *Front. Immunol.* **2019**, *10*, 790. [CrossRef]
10. Tobin, M.K.; Bonds, J.A.; Minshall, R.D.; Pelligrino, D.A.; Testai, F.D.; Lazarov, O. Neurogenesis and inflammation after ischemic stroke: What is known and where we go from here. *J. Cereb. Blood Flow Metab.* **2014**, *34*, 1573–1584. [CrossRef]
11. Boshuizen, M.C.S.; Steinberg, G.K. Stem Cell–Based Immunomodulation After Stroke: Effects on Brain Repair Processes. *Stroke* **2018**, *49*, 1563–1570. [CrossRef] [PubMed]
12. Cherry, J.D.; Olschowka, J.A.; O'Banion, M.K. Neuroinflammation and M2 microglia: The good, the bad, and the inflamed. *J. Neuroinflamm.* **2014**, *11*, 98. [CrossRef]
13. Häke, I.; Schönenberger, S.; Neumann, J.; Franke, K.; Paulsen-Merker, K.; Reymann, K.; Ismail, G.; Din, L.; Said, I.M.; Latiff, A.; et al. Neuroprotection and enhanced neurogenesis by extract from the tropical plant Knema laurina after inflammatory damage in living brain tissue. *J. Neuroimmunol.* **2009**, *206*, 91–99. [CrossRef] [PubMed]
14. Boese, A.C.; Le, Q.-S.E.; Pham, D.; Hamblin, M.H.; Lee, J.-P. Neural stem cell therapy for subacute and chronic ischemic stroke. *Stem Cell Res. Ther.* **2018**, *9*, 154. [CrossRef] [PubMed]
15. Guerra-Crespo, M.; De la Herran-Arita, A.K.; Boronat-Garcia, A.; Maya-Espinosa, G.; Garcia-Montes, J.R.; Fallon, J.H.; Drucker-Colín, R. Neural Stem Cells: Exogenous and Endogenous Promising Therapies for Stroke. *Neural Stem Cells Ther.* **2012**, 297–342. [CrossRef]
16. Ma, Y.; Wang, J.; Wang, Y.; Yang, G.-Y. The biphasic function of microglia in ischemic stroke. *Prog. Neurobiol.* **2017**, *157*, 247–272. [CrossRef]
17. Ariffin, S.H.Z.; Wahab, R.M.A.; Ismail, I.; Mahadi, N.M.; Ariffin, Z.Z. Stem cells, cytokines and their receptors. *Asia-Pac. J. Mol. Biol. Biotechnol.* **2005**, *13*, 1–13.
18. Hu, X.; Li, P.; Guo, Y.; Wang, H.; Leak, R.K.; Chen, S.; Chen, J. Microglia/Macrophage Polarization Dynamics Reveal Novel Mechanism of Injury Expansion After Focal Cerebral Ischemia. *Stroke* **2012**, *43*, 3063–3070. [CrossRef] [PubMed]
19. Kizil, C.; Kyritsis, N.; Brand, M. Effects of inflammation on stem cells: Together they strive? *EMBO Rep.* **2015**, *16*, 416–426. [CrossRef]
20. Reynolds, A.; Rubin, J.; Clermont, G.; Day, J.; Vodovotz, Y.; Ermentrout, G.B. A reduced mathematical model of the acute inflammatory response: I. Derivation of model and analysis of anti-inflammation. *J. Theor. Biol.* **2006**, *242*, 220–236. [CrossRef] [PubMed]
21. Kumar, R.; Clermont, G.; Vodovotz, Y.; Chow, C.C. The dynamics of acute inflammation. *J. Theor. Biol.* **2004**, *230*, 145–155. [CrossRef] [PubMed]
22. Ma, B.F.; Liu, X.M.; Zhang, A.X.; Wang, P.; Zhang, X.M.; Li, S.N.; Lahn, B.T.; Xiang, A.P. Mathematical models for the proliferation of neural stem/progenitor cells in clonogenic culture. *Rejuvenation Res.* **2007**, *10*, 205–214. [CrossRef]
23. Ziebell, F.; Martin-Villalba, A.; Marciniak-Czochra, A. Mathematical modelling of adult hippocampal neurogenesis: Effects of altered stem cell dynamics on cell counts and bromodeoxyuridine-labelled cells. *J. R. Soc. Interface* **2014**, *11*, 20140144. [CrossRef]
24. Cacao, E.; Cucinotta, F.A. Modeling Impaired Hippocampal Neurogenesis after Radiation Exposure. *Radiat. Res.* **2016**, *185*, 319–331. [CrossRef]
25. Nakata, Y.; Getto, P.; Marciniak-Czochra, A.; Alarcón, T. Stability analysis of multi-compartment models for cell production systems. *J. Biol Dyn.* **2012**, *6*, 2–18. [CrossRef]
26. Alharbi, S.A.; Rambely, A.S. Dynamic Simulation for Analyzing the Effects of the Intervention of Vitamins on Delaying the Growth of Tumor Cells. *IEEE Access* **2019**, *7*, 128816–128827. [CrossRef]
27. Alharbi, S.; Rambely, A. A dynamic simulation of the immune system response to inhibit and eliminate abnormal cells. *Symmetry* **2019**, *11*, 572. [CrossRef]
28. Alharbi, S.A.; Rambely, A.S. Dynamic behaviour and stabilisation to boost the immune system by complex interaction between tumour cells and vitamins intervention. *Adv. Differ. Equ.* **2020**, *1*, 1–18. [CrossRef]
29. Alharbi, S.A.; Rambely, A.S. A New ODE-Based Model for Tumor Cells and Immune System Competition. *Mathematics* **2020**, *8*, 1285. [CrossRef]
30. Stone, L.L.H.; Grande, A.W.; Low, W.C. Neural Repair and Neuroprotection with Stem Cells in Ischemic Stroke. *Brain Sci.***2013**, *3*, 599–614. [CrossRef] [PubMed]
31. Alqudah, M.A. Cancer treatment by stem cells and chemotherapy as a mathematical model with numerical simulations. *Alex. Eng. J.* **2020**, *59*, 1953–1957. [CrossRef]
32. Wei, L.; Wei, Z.Z.; Jiang, M.Q.; Mohamad, O.; Yu, S.P. Stem cell transplantation therapy for multifaceted therapeutic benefits after stroke. *Prog. Neurobiol.* **2017**, *157*, 49–78. [CrossRef]
33. Zhang, G.-L.; Zhu, Z.-H.; Wang, Y.-Z. Neural stem cell transplantation therapy for brain ischemic stroke: Review and perspectives. *World J. Stem Cells* **2019**, *11*, 817–830. [CrossRef]
34. Guruswamy, R.; ElAli, A. Complex roles of microglial cells in ischemic stroke pathobiology: New insights and future directions. *Int. J. Mol. Sci.* **2017**, *18*, 496. [CrossRef] [PubMed]

35. Wang, J.; Xing, H.; Wan, L.; Jiang, X.; Wang, C.; Wu, Y. Treatment targets for M2 microglia polarization in ischemic stroke. *Biomed. Pharmacother.* **2018**, *105*, 518–525. [CrossRef] [PubMed]
36. Kim, J.Y.; Kim, N.; Yenari, M.A. Mechanisms and potential therapeutic applications of microglial activation after brain injury. *CNS Neurosci. Ther.* **2019**, *21*, 309–319. [CrossRef] [PubMed]
37. Bliss, T.M.; Andres, R.H.; Steinberg, G.K. Optimizing the success of cell transplantation therapy for stroke. *Neurobiol. Dis.* **2010**, 37, 275–283. [CrossRef] [PubMed]
38. Hartman, P. *Ordinary Differential Equations*; John Wiley & Sons: New York, NY, USA; London, UK; Sydney, Australia, 1964.
39. Gantmacher, F.R. *The Theory of Matrices*; American Mathematical Society: Providence, RI, USA, 1959; Volume 2.
40. Liao, L.-Y.; Lau, B.W.-M.; Sánchez-Vidaña, D.I.; Gao, Q. Exogenous neural stem cell transplantation for cerebral ischemia. *Neural Regen. Res.* **2019**, *14*, 1129–1137. [PubMed]
41. Baker, E.W.; Kinder, H.A.; West, F.D. Neural stem cell therapy for stroke: A multimechanistic approach to restoring neurological function. *Brain Behav.* **2019**, *9*, 3. [CrossRef]
42. Johnston, M.D.; Edwards, C.M.; Bodmer, W.F.; Maini, P.K.; Chapman, S.J. Examples of mathematical modeling: Tales from the crypt. *Cell Cycle* **2007**, *6*, 2106–2112. [CrossRef]

Article

Second-Order Non-Canonical Neutral Differential Equations with Mixed Type: Oscillatory Behavior

Osama Moaaz [1,*,†], Amany Nabih [1,†], Hammad Alotaibi [2,†] and Y. S. Hamed [2,†]

1 Department of Mathematics, Faculty of Science, Mansoura University, Mansoura 35516, Egypt; amanynabih89@students.mans.edu.eg
2 Department of Mathematics and Statistics, College of Science, Taif University, P. O. Box 11099, Taif 21944, Saudi Arabia; hm.alotaibi@tu.edu.sa (H.A.); yasersalah@tu.edu.sa (Y.S.H.)
* Correspondence: o_moaaz@mans.edu.eg
† These authors contributed equally to this work.

Abstract: In this paper, we establish new sufficient conditions for the oscillation of solutions of a class of second-order delay differential equations with a mixed neutral term, which are under the non-canonical condition. The results obtained complement and simplify some known results in the relevant literature. Example illustrating the results is included.

Keywords: non-canonical differential equations; second-order; neutral delay; mixed type; oscillation criteria

Citation: Moaaz, O.; Nabih, A.; Alotaibi, H.; Hamed, Y.S. Second-Order Non-Canonical Neutral Differential Equations with Mixed Type: Oscillatory Behavior. *Symmetry* **2021**, *13*, 318. https://doi.org/10.3390/sym13020318

Academic Editors: Sergei D. Odintsov and Celina Si

Received: 12 January 2021
Accepted: 10 February 2021
Published: 14 February 2021

Publisher's Note: MDPI stays neutral with regard to jurisdictional claims in published maps and institutional affiliations.

1. Introduction

This paper discusses the oscillatory behavior of solutions of second-order functional differential equation with a mixed neutral term of the form

$$\left(r(l)\left[\left(y(l)+p_1(l)y(\rho_1(l))+p_2(l)y(\rho_2(l))\right)'\right]^{\gamma}\right)'+q(l)y^{\gamma}(\sigma(l))=0, \quad (1)$$

where $l \geq l_0$. Throughout this paper, we assume the following:

(C1) $\gamma \in Q^{+}_{odd} := \{a/b : a, b \in \mathbb{Z}^{+} \text{ are odd}\}$ and $r \in C([l_0,\infty),(0,\infty))$;
(C2) $\rho_1, \rho_2, \sigma \in C([l_0,\infty),\mathbb{R})$, $\rho_1(l) \leq l \leq \rho_2(l)$, $\sigma(l) \leq l$ and $\rho_1, \rho_2, \sigma \to \infty$ as $l \to \infty$;
(C3) $p_1, p_2, q \in C([l_0,\infty),[0,\infty))$ and $q(l)$ is not identically zero for large l.

Let y be a real-valued function defined for all l in a real interval $[l_y,\infty)$, $l_y \geq l_0$, and having a second derivative for all $l \in [l_y,\infty)$. The function y is called a *solution* of the differential Equation (1) on $[l_y,\infty)$ if y satisfies (1) on $[l_y,\infty)$. A nontrivial solution y of any differential equation is said to be *oscillatory* if it has arbitrary large zeros; otherwise, it is said to be *nonoscillatory*. We will consider only those solutions of (1) which exist on some half-line $[l_b,\infty)$ for $l_b \geq l_0$ and satisfy the condition $\sup\{|y(l)| : l_c \leq l < \infty\} > 0$ for any $l_c \geq l_b$.

A delay differential equation of neutral type is an equation in which the highest order derivative of the unknown function appears both with and without delay. During the last decades, there is a great interest in studying the oscillation of solutions of neutral differential equations. This is due to the fact that such equations arise from a variety of applications including population dynamics, automatic control, mixing liquids, and vibrating masses attached to an elastic bar, biology in explaining self-balancing of the human body, and in robotics in constructing biped robots, it is easy to notice the emergence of models of the neutral delay differential equations, see [1,2].

In the following, we review some of the related works that dealt with the oscillation of the neutral differential equations of mixed-type.

Grammatikopouls et al. [3] established oscillation criteria for the equation

$$\left(r(l)\psi'(l)\right)' + q(l)y(\sigma(l)) = 0, \tag{2}$$

where

$$z(l) = y(l) + p_1(l)y(l-\sigma_1) + p_2(l)y(l+\sigma_2),$$

$r(l) = 1$, $p_2(l) = 0$, $0 \leq p_1 \leq 1$, and $q(l) \geq 0$. Ruan [4] obtained some oscillation criteria for the Equation (2) by employing Riccati technique and averaging function method, when $p_2(l) = 0$ and $\sigma(l) = l - \sigma$. Arul and Shobha [5] studied the oscillatory behavior of solution of (2), when $0 \leq p_1(l) \leq p_1 < \infty$ and $0 \leq p_2(l) \leq p_2 < \infty$.

Dzurina et al. [6] presented some sufficient conditions for the oscillation of the second-order equation

$$\left(\frac{1}{r(l)}y'(l)\right)' + p(l)y(\tau(l)) + q(l)y(\sigma(l)) = 0.$$

Li [7] and Li et al. [8] studied the oscillation of solutions of the second-order equation with constant mixed arguments:

$$\left(r(l)z'(l)\right)' + q_1(l)y(l-\sigma_3) + q_2(l)y(l+\sigma_4) = 0. \tag{3}$$

Arul and Shobha [5] established some sufficient conditions for the oscillation of all solutions of Equation (3) in the canonical case, that is,

$$\int_{l_0}^{\infty} r^{-1}(\vartheta)\mathrm{d}\vartheta = \infty,$$

Thandapani et al. [9] studied the oscillation criteria for the differential equation of the form

$$(z^{\alpha}(l))'' + q(l)y^{\beta}(l-\tau_1) + p(l)y^{\gamma}(l+\tau_1) = 0.$$

Grace et al. [10] studied the oscillatory behavior of solutions of the equation

$$\left(r(l)\left(\left(y(l) + p_1(l)y^{\beta_1}(\sigma_1(l)) + p_2(l)y^{\beta_2}(\sigma_2(l))\right)'\right)^{\gamma}\right)' + q(l)y^{\gamma}(\tau(l)) = 0,$$

and considered the two cases

$$\int_{l_0}^{\infty} r^{-1/\gamma}(\vartheta)\mathrm{d}\vartheta = \infty, \tag{4}$$

and

$$\int_{l_0}^{\infty} r^{-1/\gamma}(\vartheta)\mathrm{d}\vartheta < \infty. \tag{5}$$

In [11], Tunc et al. studied the oscillatory behavior of the differential Equation (1) under the condition (4). Moreover, they considered the two following cases: $p_1(l) \geq 0$, $p_2(l) \geq 1$, and $p_2(l) \neq 1$ eventually; $p_2(l) \geq 0$, $p_1(l) \geq 1$, and $p_2(l) \neq 1$ eventually.

For the third-order equations, Han et al. [12] studied the oscillation and asymptotic properties of the third-order equation

$$\left(a(l)z''(l)\right)' + q_1(l)y(l-\tau_3) + q_2(l)y(l+\tau_4) = 0,$$

and established two theorems which guarantee that the above equation oscillates or tends to zero. Moaaz et al. [13] discussed the oscillation and asymptotic behavior of solutions of the third-order equation

$$\left(r(l)\left(x''(l)\right)^{\alpha}\right)' + q_1(l)f_1(y(\sigma_1(l))) + q_2(l)f_2(y(\sigma_2(l))) = 0,$$

where $x(l) = y(l) + p_1(l)y(\tau_1(l)) + p_2(l)y(\tau_2(l))$. For further results, techniques, and approaches in studying oscillation of the delay differential equations, see in [14–24].

In this paper, we study the oscillatory behavior of solutions of the second-order differential equation with a mixed neutral term (1) under condition (5). We follow a new approach based on deducing a new relationship between the solution and the corresponding function. Using this new relationship, we first obtain one condition that ensures oscillation of (1). Moreover, by introducing a generalized Riccati substitution, we get a new criterion for oscillation of (1). Often these types of equations (such as (1), (2), and (3)) are studied under condition (4). On the other hand, the works that studied these equations under the condition (5) obtained two conditions to ensure the oscillation. Therefore, our results are an extension and simplification as well as improvement of previous results in [3–5,8,11].

2. Main Results

We adopt the following notation for a compact presentation of our results:

$$\psi(l) := y(l) + p_1(l)y(\rho_1(l)) + p_2(l)y(\rho_2(l)),$$

$$\kappa(u,v) := \int_u^v r^{-1/\gamma}(\delta)\mathrm{d}\delta,$$

$$B_1(l) := 1 - p_1(l)\frac{\kappa(\rho_1(l),\infty)}{\kappa(l,\infty)} - p_2(l)$$

and

$$B_2(l) := 1 - p_1(l) - p_2(l)\frac{\kappa(l_1,\rho_2(l))}{\kappa(l_1,l)}.$$

Lemma 1. *Assume that* $\Theta(\vartheta) := A\vartheta - B(\vartheta - C)^{(\gamma+1)/\gamma}$*, where* A, B *and* C *are real constants;* $B > 0$*; and* $\gamma \in Q^{+}_{odd}$*. Then,*

$$\Theta(\vartheta^*) \le \max_{u\in\mathbb{R}}\Theta(\vartheta) = AC + \frac{\gamma^\gamma}{(\gamma+1)^{\gamma+1}}A^{\gamma+1}B^{-\gamma}.$$

Lemma 2. *Assume that* y *is a positive solution of (1) on* $[l_0,\infty)$*. If* ψ *is a decreasing positive function for* $l \ge l_1$ *large enough, then*

$$\left(\frac{\psi(l)}{\kappa(l,\infty)}\right)' \ge 0, \text{ for } l \ge l_1. \tag{6}$$

While if ψ *is a increasing positive function for* $l \ge l_1$*, then*

$$\left(\frac{\psi(l)}{\kappa(l_1,l)}\right)' \le 0, \text{ for } l \ge l_1. \tag{7}$$

Proof. Assume that (1) has a positive solution y on $[l_0,\infty)$. Therefore, there exists a $l_1 \ge l_0$ such that, for all $l \ge l_1$, $\psi(l) \ge y(l) > 0$ and $(r(l)(\psi'(l))^\gamma) \le 0$. From (1), we see that

$$\left(r(l)(\psi'(l))^\gamma\right)' = -q(l)y^\gamma(\sigma(l)) \le 0.$$

Obviously, ψ is either eventually decreasing or eventually increasing.
Let ψ be a decreasing function on $[l_1,\infty)$. Then, $\lim_{l\to\infty}\psi(l) < \infty$, and so

$$\psi(l) \ge -\int_l^\infty r^{-1/\gamma}(\vartheta)r^{1/\gamma}(\vartheta)\psi'(\vartheta)\mathrm{d}\vartheta \ge -\kappa(l,\infty)r^{1/\gamma}(l)\psi'(l). \tag{8}$$

Thus,

$$\left(\frac{\psi(l)}{\kappa(l,\infty)}\right)' = \frac{\kappa(l,\infty)\psi'(l) + r^{-1/\gamma}(l)\psi(l)}{(\kappa(l,\infty))^2} \geq 0.$$

Let ψ be a increasing function on $[l_1,\infty)$. Then, we obtain

$$\psi(l) \geq \int_{l_1}^{l} r^{-1/\gamma}(\vartheta) r^{1/\gamma}(\vartheta)\psi'(\vartheta)\mathrm{d}\vartheta \geq \kappa(l_1,l) r^{1/\gamma}(l)\psi'(l),$$

and so

$$\left(\frac{\psi(l)}{\kappa(l_1,l)}\right)' = \frac{\kappa(l_1,l)\psi'(l) - r^{-1/\gamma}(l)\psi(l)}{\kappa^2(l_1,l)} \leq 0.$$

Thus, the proof is complete. □

Theorem 1. *Assume that $B_2(l) \geq B_1(l) > 0$. If*

$$\limsup_{l\to\infty} \int_{l_1}^{l} \frac{1}{r^{1/\gamma}(\beta)} \left(\int_{l_1}^{\beta} q(\delta) B_1^{\gamma}(\sigma(\delta))\kappa^{\gamma}(\sigma(\delta),\infty)\mathrm{d}\delta\right)^{1/\gamma} \mathrm{d}\beta = \infty, \tag{9}$$

then, all solutions of (1) are oscillatory.

Proof. Assume the contrary that Equation (1) has a positive solution y on $[l_0,\infty)$. Then, $y(\rho_1(l))$, $y(\rho_2(l))$ and $y(\sigma(l))$ are positive for all $l \geq l_1$, where l_1 is large enough. Thus, from (1) and the definition of ψ, we note that $\psi(l) \geq y(l) > 0$ and $r(l)(\psi'(l))^{\gamma}$ is nonincreasing. Therefore, ψ' is either eventually negative or eventually positive.
Let $\psi'(l) < 0$ on $[l_1,\infty)$. By using Lemma 2, we have

$$\psi(\rho_1(l)) \leq \frac{\kappa(\rho_1(l),\infty)}{\kappa(l,\infty)}\psi(l),$$

based on the fact that $\rho_1(l) \leq l$. Therefore,

$$\begin{aligned} y(l) &= \psi(l) - p_1(l)y(\rho_1(l)) - p_2(l)y(\rho_2(l)) \\ &\geq \psi(l) - p_1(l)\psi(\rho_1(l)) - p_2(l)\psi(\rho_2(l)) \\ &\geq \left(1 - p_1(l)\frac{\kappa(\rho_1(l),\infty)}{\kappa(l,\infty)} - p_2(l)\right)\psi(l) \\ &= B_1(l)\psi(l). \end{aligned}$$

Therefore, (1) becomes

$$\left(r(l)(\psi'(l))^{\gamma}\right)' \leq -q(l)B_1^{\gamma}(\sigma(l))\psi^{\gamma}(\sigma(l)). \tag{10}$$

As $\left(r(l)(\psi'(l))^{\gamma}\right)' \leq 0$, we have

$$r(l)(\psi'(l))^{\gamma} \leq r(l_1)(\psi'(l_1))^{\gamma} := -L < 0, \tag{11}$$

for all $l \geq l_1$, from (8) and (11), we have

$$\psi^{\gamma}(l) \geq L\kappa^{\gamma}(l,\infty) \text{ for all } l \geq l_1. \tag{12}$$

Combining (10) with (12) yields

$$\left(r(l)(\psi'(l))^{\gamma}\right)' \leq -Lq(l)B_1^{\gamma}(\sigma(l))\kappa^{\gamma}(\sigma(l),\infty), \tag{13}$$

for all $l \geq l_1$. Integrating (13) from l_1 to l, we obtain

$$\begin{aligned} r(l)(\psi'(l))^\gamma &\leq r(l_1)(\psi'(l_1))^\gamma - L\int_{l_1}^{l} q(\delta)B_1^\gamma(\sigma(\delta))\kappa^\gamma(\sigma(\delta),\infty)\mathrm{d}\delta \\ &\leq -L\int_{l_1}^{l} q(\delta)B_1^\gamma(\sigma(\delta))\kappa^\gamma(\sigma(\delta),\infty)\mathrm{d}\delta. \end{aligned}$$

Integrating the last inequality from l_1 to l, we get

$$\psi(l) \leq \psi(l_1) - L^{1/\gamma}\int_{l_1}^{l} \frac{1}{r^{1/\gamma}(\beta)}\left(\int_{l_1}^{\beta} q(\delta)B_1^\gamma(\sigma(\delta))\kappa^\gamma(\sigma(\delta),\infty)\mathrm{d}\delta\right)^{1/\gamma}\mathrm{d}\beta.$$

At $l \to \infty$, we get a contradiction with (9).
Let $\psi'(l) > 0$ on $[l_1,\infty)$. From Lemma 2, we arrive at

$$\psi(\rho_2(l)) \leq \frac{\kappa(l_1,\rho_2(l))}{\kappa(l_1,l)}\psi(l). \tag{14}$$

From the definition of ψ, we obtain

$$\begin{aligned} y(l) &= \psi(l) - p_1(l)y(\rho_1(l)) - p_2(l)y(\rho_2(l)) \\ &\geq \psi(l) - p_1(l)\psi(\rho_1(l)) - p_2(l)\psi(\rho_2(l)). \end{aligned} \tag{15}$$

Using that (14) and $\psi(\rho_1(l)) \leq \psi(l)$ where $\rho_1(l) < l$ in (15), we obtain

$$\begin{aligned} y(l) &\geq \psi(l)\left(1 - p_1(l) - p_2(l)\frac{\kappa(l_1,\rho_2(l))}{\kappa(l_1,l)}\right) \\ &\geq B_2(l)\psi(l). \end{aligned} \tag{16}$$

Thus, (1) becomes

$$\left(r(l)(\psi'(l))^\gamma\right)' \leq -q(l)B_2^\gamma(\sigma(l))\psi^\gamma(\sigma(l)). \tag{17}$$

Now, from (9) and (C2), we have that $\int_{l_1}^{l} q(\vartheta)B_1^\gamma(\sigma(\vartheta))\kappa^\gamma(\sigma(\vartheta),\infty)\mathrm{d}\vartheta$ is unbounded. Therefore, as $\kappa'(l,\infty) < 0$, we obtain that

$$\int_{l_1}^{l} q(\vartheta)B_1^\gamma(\sigma(\vartheta))\mathrm{d}\vartheta \to \infty \text{ as } l \to \infty. \tag{18}$$

Integrating (17) from l_2 to l, we get

$$\begin{aligned} r(l)(\psi'(l))^\gamma &\leq r(l_2)(\psi'(l_2))^\gamma - \int_{l_2}^{l} q(\vartheta)B_2^\gamma(\sigma(\vartheta))\psi^\gamma(\sigma(\vartheta))\mathrm{d}\vartheta \\ &\leq r(l_2)(\psi'(l_2))^\gamma - \psi^\gamma(\sigma(l_2))\int_{l_2}^{l} q(\vartheta)B_2^\gamma(\sigma(\vartheta))\mathrm{d}\vartheta. \end{aligned}$$

As $B_2(l) > B_1(l)$, we get

$$r(l)(\psi'(l))^\gamma \leq r(l_2)(\psi'(l_2))^\gamma - \psi^\gamma(\sigma(l_2))\int_{l_2}^{l} q(\vartheta)B_1^\gamma(\sigma(\vartheta))\mathrm{d}\vartheta. \tag{19}$$

From (18) and (19), we get a contradiction with the positivity of $\psi'(l)$. Therefore, the proof is complete. □

Theorem 2. *Assume that* $B_2(l) \geq B_1(l) > 0$. *If*

$$\limsup_{l\to\infty} \kappa^{\gamma}(l,\infty) \int_{l_1}^{l} q(\vartheta)B_1^{\gamma}(\vartheta)\mathrm{d}\vartheta > 1, \tag{20}$$

then, all solutions of (1) are oscillatory.

Proof. Assume the contrary that Equation (1) has a positive solution y on $[l_0,\infty)$. Then, $y(\rho_1(l))$, $y(\rho_2(l))$ and $y(\sigma(l))$ are positive for all $l \geq l_1$, where l_1 is large enough. Thus, from (1) and the definition of ψ, we note that $\psi(l) \geq y(l) > 0$ and $r(l)(\psi'(l))^{\gamma}$ is nonincreasing. Therefore, ψ' is either eventually negative or eventually positive.
Let $\psi'(l) < 0$ on $[l_1,\infty)$. Integrating (10) from l_1 to l, we get

$$\begin{aligned} r(l)(\psi'(l))^{\gamma} &\leq r(l_1)(\psi'(l_1))^{\gamma} - \int_{l_1}^{l} q(\vartheta)B_1^{\gamma}(\sigma(\vartheta))\psi^{\gamma}(\sigma(\vartheta))\mathrm{d}\vartheta \\ &\leq -\psi^{\gamma}(\sigma(l)) \int_{l_1}^{l} q(\vartheta)B_1^{\gamma}(\vartheta)\mathrm{d}\vartheta. \end{aligned} \tag{21}$$

Using $\psi(\sigma(l)) \geq \psi(l)$ and (8) in (21), we obtain

$$-r(l)(\psi'(l))^{\gamma} \geq -r(l)(\psi'(l))^{\gamma}\kappa^{\gamma}(l,\infty) \int_{l_1}^{l} q(\vartheta)B_1^{\gamma}(\vartheta)\mathrm{d}\vartheta. \tag{22}$$

Divide both sides of inequality (22) by $-r(l)(\psi'(l))^{\gamma}$ and taking the limsup, we get

$$\limsup_{l\to\infty} \kappa^{\gamma}(l,\infty) \int_{l_1}^{l} q(\vartheta)B_1^{\gamma}(\vartheta)\mathrm{d}\vartheta \leq 1.$$

Thus, we get a contradiction with (20).

Let $\psi' > 0$ on $[l_1,\infty)$. From (20) and the fact that $\kappa(l,\infty) < \infty$, we have that (18) holds. Then, this part of proof is similar to that of Theorem 1. Therefore, the proof is complete. □

Theorem 3. *Assume that* $B_2(l) > 0$, $B_1(l) > 0$ *and* $r' > 0$. *If there exist positive functions* $\mu, \delta \in C^1([l_0,\infty))$ *and* $l_1 \in [l_0,\infty)$ *such that*

$$\limsup_{l\to\infty} \left\{ \frac{\kappa^{\gamma}(l,\infty)}{\delta(l)} \int_{l_1}^{l} \left(\delta(\vartheta)q(\vartheta)B_1^{\gamma}(\sigma(\vartheta)) - \frac{r(\vartheta)}{(\gamma+1)^{\gamma+1}} \frac{(\delta'(\vartheta))^{\gamma+1}}{(\delta(\vartheta))^{\gamma}} \right) \mathrm{d}\vartheta \right\} > 1 \tag{23}$$

and

$$\limsup_{l\to\infty} \int_{l_1}^{l} \left(\mu(\vartheta)q(\vartheta)B_2^{\gamma}(\sigma(\vartheta)) - \frac{1}{(\gamma+1)^{\gamma+1}} \frac{r(\vartheta)(\mu'(\vartheta))^{\gamma+1}}{\mu^{\gamma}(\vartheta)(\sigma'(\vartheta))^{\gamma}} \right) \mathrm{d}\vartheta = \infty, \tag{24}$$

then, all solutions of (1) are oscillatory.

Proof. Assume the contrary that Equation (1) has a positive solution y on $[l_0,\infty)$. Then, $y(\rho_1(l))$, $y(\rho_2(l))$ and $y(\sigma(l))$ are positive for all $l \geq l_1$, where l_1 is large enough. Thus, from (1) and the definition of ψ, we note that $\psi(l) \geq y(l) > 0$ and $r(l)(\psi'(l))^{\gamma}$ is nonincreasing. Therefore, ψ' is either eventually negative or eventually positive.
Let $\psi' < 0$ on $[l_1,\infty)$. As in proof of Theorem 1, we arrive at (10). Now, we define the function

$$\omega(l) = \delta(l)\left(\frac{r(l)(\psi'(l))^{\gamma}}{\psi^{\gamma}(l)} + \frac{1}{\kappa^{\gamma}(l,\infty)} \right) \text{ on } [l_1,\infty). \tag{25}$$

From (8), we have that $\omega \geq 0$ on $[l_1, \infty)$. Differentiating (25), we get

$$\begin{aligned}
\omega'(l) &= \frac{\delta'(l)}{\delta(l)}\omega(l) + \delta(l)\frac{\left(r(l)(\psi'(l))^{\gamma}\right)'}{\psi^{\gamma}(l)} - \gamma\delta(l)r(l)\left(\frac{\psi'(l)}{\psi(l)}\right)^{\gamma+1} \\
&\quad + \frac{\gamma\delta(l)}{r^{1/\gamma}(l)\kappa^{\gamma+1}(l,\infty)} \\
&\leq \frac{\delta'(l)}{\delta(l)}\omega(l) + \delta(l)\frac{\left(r(l)(\psi'(l))^{\gamma}\right)'}{\psi^{\gamma}(l)} - \frac{\gamma}{(\delta(l)r(l))^{1/\gamma}}\left(\omega(l) - \frac{\delta(l)}{\kappa^{\gamma}(l,\infty)}\right)^{(\gamma+1)/\gamma} \\
&\quad + \frac{\gamma\delta(l)}{r^{1/\gamma}(l)\kappa^{\gamma+1}(l,\infty)}. \qquad (26)
\end{aligned}$$

Combining (10) and (26), we have

$$\begin{aligned}
\omega'(l) &\leq -\frac{\gamma}{(\delta(l)r(l))^{1/\gamma}}\left(\omega(l) - \frac{\delta(l)}{\kappa^{\gamma}(l,\infty)}\right)^{(\gamma+1)/\gamma} - \delta(l)q(l)B_1^{\gamma}(\sigma(l))\frac{\psi^{\gamma}(\sigma(l))}{\psi^{\gamma}(l)} \\
&\quad + \frac{\gamma\delta(l)}{r^{1/\gamma}(l)\kappa^{\gamma+1}(l,\infty)} + \frac{\delta'(l)}{\delta(l)}\omega(l). \qquad (27)
\end{aligned}$$

Using Lemma 1 with $A := \delta'(l)/\delta(l)$, $B := \gamma(\delta(l)r(l))^{-1/\gamma}$, $C := \delta(l)/\kappa^{\gamma}(l,\infty)$ and $\vartheta := \omega$, we get

$$\begin{aligned}
\frac{\delta'(l)}{\delta(l)}\omega(l) - \frac{\gamma}{(\delta(l)r(l))^{1/\gamma}}\left(\omega(l) - \frac{\delta(l)}{\kappa^{\gamma}(l,\infty)}\right)^{(\gamma+1)/\gamma} &\leq \frac{1}{(\gamma+1)^{\gamma+1}}r(l)\frac{(\delta'(l))^{\gamma+1}}{(\delta(l))^{\gamma}} \\
&\quad + \frac{\delta'(l)}{\kappa^{\gamma}(l,\infty)}.
\end{aligned}$$

As $l \geq \sigma(l)$, we arrive at

$$\psi(\sigma(l)) \geq \psi(l), \qquad (28)$$

which, in view of (27) and (28), gives

$$\begin{aligned}
\omega'(l) &\leq \frac{\delta'(l)}{\kappa^{\gamma}(l,\infty)} + \frac{1}{(\gamma+1)^{\gamma+1}}r(l)\frac{(\delta'(l))^{\gamma+1}}{(\delta(l))^{\gamma}} - \delta(l)q(l)B_1^{\gamma}(\sigma(l))\frac{\psi^{\gamma}(\sigma(l))}{\psi^{\gamma}(l)} \\
&\quad + \frac{\gamma\delta(l)}{r^{1/\gamma}(l)\kappa^{\gamma+1}(l,\infty)} \\
&\leq -\delta(l)q(l)B_1^{\gamma}(\sigma(l)) + \left(\frac{\delta(l)}{\kappa^{\gamma}(l,\infty)}\right)' + \frac{r(l)}{(\gamma+1)^{\gamma+1}}\frac{(\delta'(l))^{\gamma+1}}{(\delta(l))^{\gamma}}. \qquad (29)
\end{aligned}$$

Integrating (29) from l_2 to l, we arrive at

$$\begin{aligned}
\int_{l_2}^{l}\left(\delta(\vartheta)q(\vartheta)B_1^{\gamma}(\sigma(\vartheta)) - \frac{r(\vartheta)}{(\gamma+1)^{\gamma+1}}\frac{(\delta'(\vartheta))^{\gamma+1}}{(\delta(\vartheta))^{\gamma}}\right)d\vartheta &\leq \left(\frac{\delta(l)}{\kappa^{\gamma}(l,\infty)} - \omega(l)\right)_{l_2}^{l} \\
&\leq -\left(\delta(l)\frac{r(l)(\psi'(l))^{\gamma}}{\psi^{\gamma}(l)}\right)_{l_2}^{l}. \qquad (30)
\end{aligned}$$

From (8), we have

$$-\frac{r^{1/\gamma}(l)\psi'(l)}{\psi(l)} \leq \frac{1}{\kappa(l,\infty)},$$

which, in view of (30), implies

$$\frac{\kappa^{\gamma}(l,\infty)}{\delta(l)}\int_{l_2}^{l}\left(\delta(\vartheta)q(\vartheta)B_1^{\gamma}(\sigma(\vartheta)) - \frac{r(\vartheta)}{(\gamma+1)^{\gamma+1}}\frac{(\delta'(\vartheta))^{\gamma+1}}{(\delta(\vartheta))^{\gamma}}\right)d\vartheta \leq 1.$$

Thus, we get a contradiction with (23).

Let $\psi'(l) > 0$ on $[l_1, \infty)$. As in proof of Theorem 1, we arrive at (17). Now, we define the function

$$\varphi(l) = \mu(l)\frac{r(l)(\psi'(l))^{\gamma}}{\psi^{\gamma}(\sigma(l))}. \tag{31}$$

Therefore, we have that $\omega \geq 0$ on $[l_1, \infty)$. Differentiating (31), we find

$$\varphi'(l) = \frac{\mu'(l)}{\mu(l)}\varphi(l) + \mu(l)\frac{\left(r(l)(\psi'(l))^{\gamma}\right)'}{\psi^{\gamma}(\sigma(l))} - \gamma\mu(l)r(l)\frac{(\psi'(l))^{\gamma}\psi'(\sigma(l))\sigma'(l)}{\psi^{\gamma+1}(\sigma(l))}. \tag{32}$$

Combining (17) and (32), we have

$$\varphi'(l) \leq \frac{\mu'(l)}{\mu(l)}\varphi(l) - \mu(l)q(l)B_2^{\gamma}(\sigma(l)) - \gamma\mu(l)r(l)\frac{(\psi'(l))^{\gamma}\psi'(\sigma(l))\sigma'(l)}{\psi^{\gamma+1}(\sigma(l))}.$$

As $\left(r(l)(\psi'(l))^{\gamma}\right)' < 0$ and $\sigma(l) \leq l$, we arrive at

$$\varphi'(l) \leq \frac{\mu'(l)}{\mu(l)}\varphi(l) - \mu(l)q(l)B_2^{\gamma}(\sigma(l)) - \gamma\mu(l)r(l)\sigma'(l)\frac{(\psi'(l))^{\gamma+1}}{\psi^{\gamma+1}(\sigma(l))}.$$

From (31), we have

$$\varphi'(l) \leq \frac{\mu'(l)}{\mu(l)}\varphi(l) - \mu(l)q(l)B_2^{\gamma}(\sigma(l)) - \frac{\gamma\sigma'(l)}{\mu^{1/\gamma}(l)r^{1/\gamma}(l)}\phi^{(\gamma+1)/\gamma}(l).$$

Using the inequality

$$Kv - Lv^{(\gamma+1)/\gamma} \leq \frac{\gamma^{\gamma}}{(\gamma+1)^{\gamma+1}}\frac{K^{\gamma+1}}{L^{\gamma}},\ L > 0, \tag{33}$$

with $K = \mu'(l)/\mu(l)$, $L = \gamma\sigma'(l)/\mu^{1/\gamma}(l)r^{1/\gamma}(l)$ and $v = \varphi$, we have

$$\varphi'(l) \leq -\mu(l)q(l)B_2^{\gamma}(\sigma(l)) + \frac{1}{(\gamma+1)^{\gamma+1}}\frac{r(l)(\mu'(l))^{\gamma+1}}{\mu^{\gamma}(l)(\sigma'(l))^{\gamma}}. \tag{34}$$

Integrating (34) from l_2 to l, we arrive at

$$\int_{l_2}^{l}\left(\mu(\vartheta)q(\vartheta)B_2^{\gamma}(\sigma(\vartheta)) - \frac{1}{(\gamma+1)^{\gamma+1}}\frac{r(\vartheta)(\mu'(\vartheta))^{\gamma+1}}{\mu^{\gamma}(\vartheta)(\sigma'(\vartheta))^{\gamma}}\right)\mathrm{d}\vartheta \leq \varphi(l_2).$$

Taking the lim sup on both sides of this inequality, we have a contradiction with (24). The proof of the theorem is complete. □

Example 1. *Consider the second-order neutral differential equation*

$$\left(l^2\left(\left(y(l) + p_0 y\left(\frac{l}{\lambda}\right) + p_* y(\lambda l)\right)'\right)\right)' + q_0 l y(\sigma_0 l) = 0, \tag{35}$$

where $\lambda > 1$, $\sigma_0 \in (0,1)$ *and* $(\lambda p_0 + p_*) \in (0,1)$. *We note that* $r(l) = l^2$, $p_1(l) = p_0$, $p_2(l) = p_*$, $\rho_1(l) = l/\lambda$, $\rho_2(l) = \lambda l$, $q(l) = q_0 l$ *and* $\sigma(l) = \sigma_0 l$. *It is easy to verify that*

$$B_1(l) = 1 - \lambda p_0 - p_*,$$

and

$$B_2(l) = 1 - p_0 - p_* \left(\frac{l - \frac{1}{\lambda}}{l - 1} \right),$$

and so $B_2 > B_1 > 0$. *Now, we see that*

$$\limsup_{l \to \infty} \int_{l_1}^{l} \frac{1}{r^{1/\gamma}(\beta)} \left(\int_{l_1}^{\beta} q(\delta) B_1^{\gamma}(\sigma(\delta)) \kappa^{\gamma}(\sigma(\delta), \infty) \mathrm{d}\delta \right)^{1/\gamma} \mathrm{d}\beta$$
$$= \limsup_{l \to \infty} \int_{l_1}^{l} \frac{1}{\beta^2} \left(\int_{l_1}^{\beta} q_0 \delta (1 - \lambda p_0 - p_*) \frac{1}{\sigma_0 \delta} \mathrm{d}\delta \right) \mathrm{d}\beta = \infty.$$

Then, by Theorem 1, we have that (35) is oscillatory.

3. Conclusions

In this work, new criteria to test the oscillation of the solutions of second-order non-canonical neutral differential equations with mixed type were presented. These criteria are to further complement and simplify relevant results in the literature.

Author Contributions: Conceptualization, O.M. and Y.S.H.; methodology, H.A.; investigation, O.M. and A.N.; writing—original draft preparation, O.M., A.N. and Y.S.H.; writing—review and editing, A.N. and H.A. All authors have read and agreed to the published version of the manuscript.

Funding: There is no external funding for this article.

Institutional Review Board Statement: Not applicable.

Informed Consent Statement: Not applicable.

Data Availability Statement: Not applicable.

Acknowledgments: This research was supported by Taif University Researchers Supporting Project Number (TURSP-2020/155), Taif University, Taif, Saudi Arabia.

Conflicts of Interest: The authors declare no conflicts of interest.

References

1. Hale, J.K. Partial neutral functional differential equations. *Rev. Roum. Math. Pures Appl.* **1994**, *39*, 339–344.
2. MacDonald, N. *Biological Delay Systems: Linear Stability Theory*; Cambridge University Press: Cambridge, UK, 1989.
3. Grammatikopouls, M.K.; Ladas, G.; Meimaridou, A. Oscillation of second order neutral delay differential equations. *Rad. Math.* **1985**, *1*, 267–274.
4. Ruan, S.G. Oscillations of second order neutral differential equations. *Can. Math. Bull.* **1993**, *36*, 485–496. [CrossRef]
5. Arul, R.; Shobha, V.S. Oscillation of second order neutral differential equations with mixed neutral term. *Int. J. Pure Appl. Math.* **2015**, *104*, 181–191. [CrossRef]
6. Dzurina, J.; Busha, J.; Airyan, E.A. Oscillation criteria for second-order differential equations of neutral type with mixed arguments. *Differ. Equ.* **2002**, *38*, 137–140. [CrossRef]
7. Li, T. Comparison theorems for second-order neutral differential equations of mixed type. *Electron. J. Differ. Equ.* **2020**, *2010*, 1–7.
8. Li, T.; Baculíková, B.; Džurina, J. Oscillation results for second-order neutral differential equations of mixed type. *Tatra Mt. Math. Publ.* **2011**, *48*, 101–116. [CrossRef]
9. Thandapani, E.; Selvarangam, S.; Vijaya, M.; Rama, R. Oscillation results for second order nonlinear differential equation with delay and advanced arguments. *Kyungpook Math. J.* **2016**, *56*, 137–146. [CrossRef]
10. Grace, S.R.; Graef, J.R.; Jadlovská, I. Oscillation criteria for second-order half-linear delay differential equations with mixed neutral terms. *Math. Slovaca* **2019**, *69*, 1117–1126. [CrossRef]
11. Tunc, E.; Ozdemir, O. On the oscillation of second-order half-linear functional differential equations with mixed neutral term. *J. Taibah Univ. Sci.* **2019**, *13*, 481–489. [CrossRef]
12. Han, Z.; Li, T.; Zhang, C.; Sun, S. Oscillatory behavior of solutions of certain third-order mixed neutral functional differential equations. *Bull. Malays. Math. Sci. Soc.* **2012**, *35*, 611–620.
13. Moaaz, O.; Chalishajar, D.; Bazighifan, O. Asymptotic behavior of solutions of the third order nonlinear mixed type neutral differential equations. *Mathematics* **2020**, *8*, 485. [CrossRef]
14. Agarwal, R.; Shieh, S.L.; Yeh, C.C. Oscillation criteria for second order retard differential equations. *Math. Comput. Model.* **1997**, *26*, 1–11. [CrossRef]

15. Agarwal, R.P.; Zhang, C.; Li, T. Some remarks on oscillation of second order neutral differential equations. *Appl. Math. Compt.* **2016**, *274*, 178–181. [CrossRef]
16. Baculikova, B.; Dzurina, J. Oscillation theorems for second-order nonlinear neutral differential equations. *Comput. Math. Appl.* **2011**, *62*, 4472–4478.
17. Bohner, M.; Grace, S.R.; Jadlovska, I. Oscillation criteria for second-order neutral delay differential equations. *Electron. J. Qual. Theory Differ. Equ.* **2017**, *2017*, 1–12. [CrossRef]
18. Chatzarakis, G.E.; Dzurina, J.; Jadlovska, I. New oscillation criteria for second-order half-linear advanced differential equations. *Appl. Math. Comput.* **2019**, *347*, 404–416. [CrossRef]
19. Moaaz, O.; Elabbasy, E.M.; Qaraad, B. An improved approach for studying oscillation of generalized Emden—Fowler neutral differential equation. *J. Inequal. Appl.* **2020**, *2020*, 69. [CrossRef]
20. Moaaz, O.; Muhib, A. New oscillation criteria for nonlinear delay differential equations of fourth-order. *Appl. Math. Comput.* **2020**, *377*, 125192. [CrossRef]
21. Sun, Y.G.; Meng, F.W. Note on the paper of Dzurina and Stavroulakis: "Oscillation criteria for second-order delay differential equations" [*Appl. Math. Comput.* **2003**, *140*, 445–453]. *Appl. Math. Comput.* **2006**, *174*, 1634–1641.
22. Xu, R.; Meng, F. Some new oscillation criteria for second order quasi-linear neutral delay differential equations. *Appl. Math. Comput.* **2006**, *182*, 797–803. [CrossRef]
23. Zhang, C.; Agarwal, R.P.; Bohner, M.; Li, T. New results for oscillatory behavior of even-order half-linear delay differential equations. *Appl. Math. Lett.* **2013**, *26*, 179–183. [CrossRef]
24. Zhang, C.; Li, T.; Suna, B.; Thandapani, E. On the oscillation of higher-order half-linear delay differential equations. *Appl. Math. Lett.* **2011**, *24*, 1618–1621. [CrossRef]

Article

Generalized Attracting Horseshoe in the Rössler Attractor

Karthik Murthy [1], Ian Jordan [2], Parth Sojitra [3], Aminur Rahman [4,*] and Denis Blackmore [5]

[1] Department of Computer Science, University of Illinois at Urbana-Champaign, Champaign, IL 61801, USA; kmurthy3@illinois.edu
[2] Department of Applied Mathematics and Statistics, Stony Brook University, Stony Brook, NY 11794, USA; ian.jordan@stonybrook.edu
[3] Department of Electrical and Computer Engineering, New Jersey Institute of Technology, Newark, NJ 07102, USA; pbs26@njit.edu
[4] Department of Applied Mathematics, University of Washington, Seattle, WA 98195, USA
[5] Department of Mathematical Sciences, New Jersey Institute of Technology, Newark, NJ 07102, USA; denis.l.blackmore@njit.edu
* Correspondence: arahman2@uw.edu

Abstract: We show that there is a mildly nonlinear three-dimensional system of ordinary differential equations—realizable by a rather simple electronic circuit—capable of producing a generalized attracting horseshoe map. A system specifically designed to have a Poincaré section yielding the desired map is described, but not pursued due to its complexity, which makes the construction of a circuit realization exceedingly difficult. Instead, the generalized attracting horseshoe and its trapping region is obtained by using a carefully chosen Poincaré map of the Rössler attractor. Novel numerical techniques are employed to iterate the map of the trapping region to approximate the chaotic strange attractor contained in the generalized attracting horseshoe, and an electronic circuit is constructed to produce the map. Several potential applications of the idea of a generalized attracting horseshoe and a physical electronic circuit realization are proposed.

Keywords: generalized attracting horseshoe; strange attractors; poincaré map; electronic circuits

Citation: Murthy, K.; Jordan, I.; Sojitra, P.; Rahman, A.; Blackmore, D. Generalized Attracting Horseshoe in the Rössler Attracto. *Symmetry* **2021**, *13*, 30. https://dx.doi.org/10.3390/sym13010030

Received: 8 December 2020
Accepted: 21 December 2020
Published: 27 December 2020

Publisher's Note: MDPI stays neutral with regard to jurisdictional claims in published maps and institutional affiliations.

1. Introduction

The seminal work of Smale [1] showed that the existence of a *horseshoe* structure in the iterate space of a diffeomorphism is enough to prove it is chaotic. Often these diffeomorphisms arise from certain Poincaré maps of continuous-time *chaotic strange attractors* (CSA), which in turn are discrete-time CSAs. Some examples of such attractors are the Lorenz strange attractor [2], the Rössler attractor [3], and the double scroll attractor [4]. An example of a Poincaré map of the Lorenz equations is the Hénon map [5], which can be further simplified to the Lozi map [6]. Unsurprisingly, symmetry (and symmetry breaking) plays an important role in the analysis of these models.

In more recent years Joshi and Blackmore [7] developed an *attracting horseshoe* (AH) model for CSAs, which has two saddles and a sink. This, however, negates the possibility of the Hénon and Lozi maps, which have two saddles. Fortunately, the attracting horseshoe can be modified into a *generalized attracting horseshoe* (GAH), which can have either one or two saddles while still being an attracting horseshoe [8]. This results in a quadrilateral trapping region. While extensive analysis was done in Joshi et al. [8], a simple concrete example seemed to be illusive.

In this investigation, we implement novel numerical techniques to find the necessary Poincaré map of the Rössler attractor that would admit a quadrilateral trapping region. This trapping region represents a region of rotational symmetry as every iterate originating in the trapping region will return after a 2π rotation. The trapping region would also filter out any flow that does not obey this symmetry. An electronic circuit could use these properties to isolate signals of interest. Similar to the experiments of Rahman et al. [9], we design a physical realization of the dynamical system in the form of an electronic circuit.

The remainder of the paper is organized as follows. In Section 2, we give an overview of the algorithm with the MATLAB codes included in the Supplementary Materials. Once we have the tools for our numerical experiments, we first propose a carefully constructed GAH model in Section 3. Then, we give numerical examples of Poincaré maps of the Rössler attractor and the map of interest in Section 4 and real-world examples in Section 5. Finally, we end discussions in Section 6 with some concluding remarks.

2. Poincaré Map Algorithm

To produce a general Poincaré section of a flow, we break up the program into four parts: solving the ODE; computing a Poincaré section perpendicular to either $x = 0, y = 0$, or $z = 0$; rotating the Poincaré section; and iterating the Poincaré map. Solving the ODE is standard through ODE45 on MATLAB, which executes a modified Runge–Kutta scheme. Once we have our solution matrix, we need to approximate the values of first return maps from the discretized flow. By restricting the first return onto a Poincaré section, the iterate space of the Poincaré map can be visualized. This is easily done for a section perpendicular to the axes, but in order to locate a highly specialized object, such as the GAH, we need to be able to rotate the section. Once the desired section is found we can experiment on iterating the points of trapping region candidates.

The initial major task is approximating the first return map on a Poincaré section of a flow. Much of the ideas of our initial first return map code came from that of Gonze [10]. Once the discretized flow is found numerically a planar section for a certain value of x, y, or z can be defined, which in general will lie between pairs of simultaneous points. Then we may draw a line between the pair through the planar section and identify the intersecting point, which approximates a point of the first return map. This can also be done with more simultaneous points in order to get higher order approximations.

Once we can approximate a map for a section perpendicular to the axes we need to have the ability to rotate and move the map to any position. This is where our program completely diverges from that in [10]. While the first instinct might be to try to rotate the section, it is equivalent to rotate the flow in the opposite direction to the desired rotation of the section. Once the flow is rotated, the code for the first return map can be readily used. This gives us the ability to analyze the first return map of a general Poincaré section.

Finally, we would like to not only compute a first return map, but also compute the iterates of a Poincaré map of any system; that is, given an initial condition on an arbitrary Poincaré section can we find the subsequent iterates. To accomplish this, we solve the ODE for a given initial condition on the planar section to find the first return. Once we have the first return, we record its location and use that as the new initial condition. This iterates the map for as many returns as desired, thereby filling in a Poincaré map. Now we have the tools needed to run numerical experiments on GAHs.

3. A Constructed GAH System

In this section, we give a brief description of the generalized attracting horseshoe (GAH) map and devise a three-dimensional nonlinear ordinary differential equation with a Poincaré section that produces it.

3.1. The GAH Map

The GAH is a modification of the AH that can be represented as a geometric paradigm with either just one or two fixed points, both of which are saddles. Figure 1 shows a rendering of a C^1 GAH with two saddle points, which can be constructed as follows. The rectangle is first contracted vertically by a factor $0 < \lambda_v < 1/2$, then expanded horizontally by a factor $1 < \lambda_h < 2$, and finally folded back into the usual horseshoe shape in such a manner that the total height and width of the horseshoe do not exceed the height and width, respectively, of the trapping rectangle Q. Then, the horseshoe is translated horizontally so that it is completely contained in Q. Obviously, the map f defined by this construction is a smooth diffeomorphism. Clearly, there are also many other ways to obtain this geometrical configuration. For example, the map f as described above is orientation-preserving, and an orientation-reversing variant can be obtained by composing it with a reflection in the horizontal axis of symmetry of the rectangle, or by composing it with a reflection in the vertical axis of symmetry followed by a composition with a half-turn. Another construction method is to use the standard Smale horseshoe that starts with a rectangle, followed by a horizontal composition with just the right scale factor or factors to move the image of Q into Q, while preserving the expansion and contraction of the horseshoe along its length and width, respectively.

It is important to note that subrectangle S with its left vertical edge through p, which contains the arch of the horseshoe and the keystone region K, plays a key role in the dynamics of the iterates of f. In particular, we require that the map satisfy the following additional property, which is illustrated in Figure 2:

($\star$) *f maps the keystone region K (containing a portion of the arch of the horseshoe) to the left of the fixed point p and the portion of its corresponding stable manifold $W^s(p)$ containing p and contained in $f(Q)$.*

The definition above and ($\star$) can be shown to lead to the conclusion that

$$\mathfrak{A} = \overline{W^u(p)} = \bigcap_{n=1}^{\infty} f^n(Q),$$

where $W^u(p)$ is the unstable manifold of p, a global chaotic strange attractor (CSA).

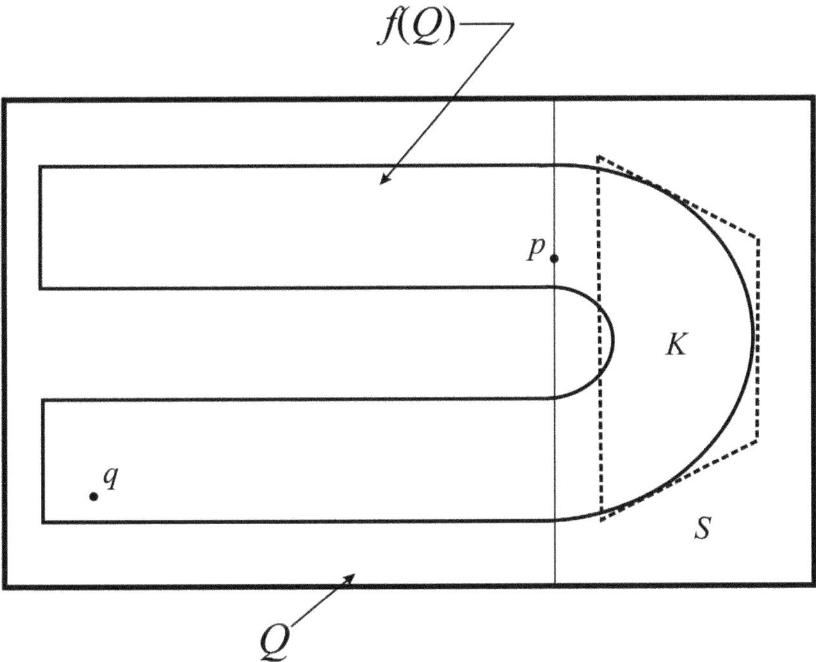

Figure 1. A planar GAH with two saddle points.

The map above can be considered to be the paradigm for a GAH, but there are many analogs. In fact, let $F : \tilde{Q} \to \tilde{Q}$ be any smooth diffeomorphism of a quadrilateral trapping region $\tilde{Q}$ possessing a horseshoe-like image with a keystone region $\tilde{K}$ containing a portion of the arch of $F(\tilde{Q})$ analogous to that shown in Figure 1. Suppose that the map is expanding by a scale factor uniformly greater than one along the length of the horseshoe and contracting transverse to it by a scale factor uniformly less than one-half in the complement of a subset of $\tilde{Q}$ containing $\tilde{K}$. Then, if F satisfies an additional property analogous to (★), it maps $\tilde{K}$ into an open subset of $\tilde{Q}$ to the left of the saddle point $\tilde{p}$, and

$$\mathfrak{A} = \overline{W^u(\tilde{p})} = \bigcap_{n=1}^{\infty} F^n(\tilde{Q})$$

is a global CSA.

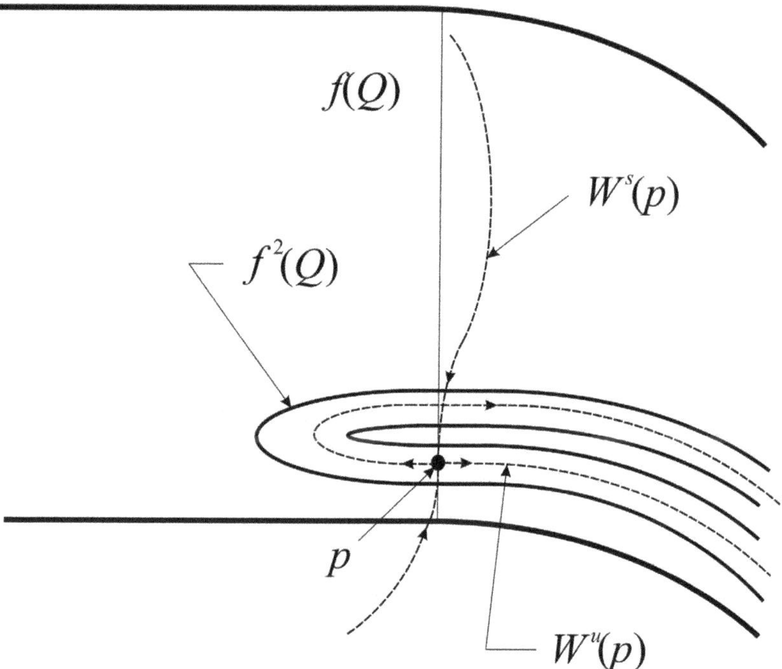

Figure 2. Local (transverse) horseshoe structure of f^2 near p.

3.2. *A GAH Producing System*

We now construct an ODE in $\mathbb{R}^3$ with a Poincaré section that is a GAH. The transversal we use is the following square in the xz-plane defined in Cartesian and polar coordinates

$$\begin{aligned} Q_0 &= \{(x,y,z) : 0.05 \leqslant x \leqslant 1.05, y = 0, -0.5 \leqslant z \leqslant 0.5\} \\ &= \{(r,\theta,z) : 0.05 \leqslant r \leqslant 1.05, \theta = 0, -0.5 \leqslant z \leqslant 0.5\}. \end{aligned} \quad (1)$$

The trick is to find a relatively simple (necessarily nonlinear) C^1 ODE having Q_0 as a transversal with an induced Poincaré first-return map $P : Q_0 \to Q_0 \supset Q_{2\pi} = P(Q_0)$ such that $P(Q_0) \subset \mathrm{int}Q_0$ is a GAH. We chose the ODE based upon a rotation about the z-axis so that the square evolves into the GAH as Q_0 makes a full rotation. The first half of the metamorphosis takes care of the vertical squeezing and horizontal stretching, while the second half produces the folding. It is not difficult to show that the system (in cylindrical coordinates)

$$\dot{r} = \frac{2\log(1.2)\sin^2\theta}{\pi}r, \ \dot{\theta} = 1, \ \dot{z} = \frac{2\log(0.2)\sin^2\theta}{\pi}(z+0.2) \quad (2)$$

flows Q_0 to

$$Q_\pi = \{(x,y,z) : -1.26 \leqslant x \leqslant -0.06, y = 0, -0.26 \leqslant z \leqslant -0.06\} \quad (3)$$

which is the original square in the radial half-plane corresponding to $\theta = 0$ stretched by a factor of 1.2 along the x-axis and squeezed by a factor of $1/5$ with respect to $z = -0.2$ along the z-axis in the radial half plane corresponding to $\theta = \pi$. Consequently, (2) produces the first half of the desired result comprising the stretching and squeezing for $0 \leqslant \theta \leqslant \pi$.

Note that (2) can be integrated directly to obtain the following for $0 \leqslant \theta \leqslant \pi$ and initial condition $(r(0), \theta(0), z(0)) = (r_0, \theta_0, z_0)$:

$$\begin{aligned} r(t) &= r(\theta) = r_0 \exp\left[\frac{\log(1.2)}{2\pi}(2\theta - \sin 2\theta)\right], \\ \theta(t) &= t, \\ z(t) &= z(\theta) = -0.2 + (z_0 + 0.2)\exp\left[\frac{\log(0.2)}{2\pi}(2\theta - \sin 2\theta)\right]. \end{aligned} \tag{4}$$

Now, we have to attend to the folding for $\pi \leqslant \theta \leqslant 2\pi$. For this we use a rotation in planes orthogonal to a fixed circle in the xy-plane. In these planes corresponding to a circle of radius c, given as $c = 0.66$, we define Euclidean coordinates with origin $r = 0.66, z = 0$ and corresponding polar coordinates (ρ, ϕ) as

$$\rho = \sqrt{(r - 0.66)^2 + z^2} = \sqrt{\tilde{r}^2 + z^2}, \tag{5}$$

where $\tilde{r} = r - 0.66 = \rho\cos\phi$ and $z = \rho\sin\phi$. Then, when $\pi \leqslant \theta \leqslant 2\pi$, we take the folding part for $\phi \geqslant -\pi/2$ to be

$$\dot{\tilde{r}} = \dot{r} = -2\sin^2\theta\,\rho\sin\phi,\ \dot{\theta} = 1,\ \dot{z} = 2\sin^2\theta\,\rho\cos\phi, \tag{6}$$

or equivalently

$$\dot{\tilde{r}} = \dot{r} = -2\sin^2\theta\, z,\ \dot{\theta} = 1,\ \dot{z} = 2\sin^2\theta\,\tilde{r}. \tag{7}$$

It is easy to verify from the above that ρ is constant (call it ρ_0) for the solutions of (6) or (7) and that the solution initially (at $t = \theta = \pi$) satisfying $(\rho, \phi) = (\rho_0, \phi_0)$ is

$$\tilde{r} = \tilde{r}(t) = \rho_0\cos(\phi(t) + \phi_0),\ \theta = \theta(t) = t,\ z = z(t) = \rho_0\sin(\phi(t) + \phi_0), \tag{8}$$

where

$$\phi(t) = (t - \pi) - \sin t\cos t. \tag{9}$$

The above ((6) or (7)) describes the folding field for $\pi \leqslant \theta \leqslant 2\pi$ and $-\pi/2 \leqslant \phi$. In order to smoothly fill in the rest of the field, we shall use the function

$$\psi(\tilde{r}) = \begin{cases} 0, & \tilde{r} \leqslant -0.6 \\ \sin^2\left[\frac{\pi}{1.2}(\tilde{r} + 0.6)\right], & -0.6 \leqslant \tilde{r} \leqslant 0 \end{cases}, \tag{10}$$

which can be recast as

$$\xi(r) = \begin{cases} 0, & r \leqslant 0.06 \\ \sin^2\left[\frac{\pi}{1.2}(r - 0.06)\right], & 0.06 \leqslant r \leqslant 0.66 \end{cases}. \tag{11}$$

We have now assembled all the elements for defining an ODE that generates a GAH Poincaré section. This ODE, which incorporates (2) and (7) and is π-periodic in θ, has the following form:

$$\dot{r} = R(r, \theta, z),\ \dot{\theta} = 1,\ \dot{z} = Z(r, \theta, z), \tag{12}$$

subject to the initial condition

$$(r(0), \theta(0), z(0)) = (r_0, 0, z_0) \in Q_0, \tag{13}$$

where

$$R = \begin{cases} \frac{\log(1.2)\sigma(\theta)r}{\pi}, & 0 \leqslant \theta \leqslant \pi \\ -\sigma(\theta)z, & (\pi \leqslant \theta \leqslant 2\pi) \text{ and } (((r \geqslant 0.66) \text{ or } (z \geqslant 0)) = (-\pi/2 \leqslant \phi \leqslant \pi)) \\ -\xi(r)\sigma(\theta)z, & (\pi \leqslant \theta \leqslant 2\pi) \text{ and } ((r < 0.66) \text{ and } (z \in [-0.26, -0.06])) \\ 0, & (\pi \leqslant \theta \leqslant 2\pi) \text{ and } ((r < 0.66) \text{ and } (z < 0) \text{ and } (z \notin [-0.26, -0.06])) \end{cases},$$

$$Z = \begin{cases} \frac{(\log(.2)\sigma(\theta)(z+0.2)}{\pi}, & 0 \leqslant \theta \leqslant \pi \\ \sigma(\theta)\tilde{r}, & (\pi \leqslant \theta \leqslant 2\pi) \text{ and } (((r \geqslant 0.66) \text{ or } (z \geqslant 0)) = (-\pi/2 \leqslant \phi \leqslant \pi)) \\ 0, & (\pi \leqslant \theta \leqslant 2\pi) \text{ and } (((r < 0.66) \text{ and } (z < 0)) = (-\pi \leqslant \phi \leqslant -\pi/2)) \end{cases},$$

and

$$\sigma(\theta) = 2\sin^2\theta = 1 - \cos 2\theta.$$

Finally, it is not difficult to show that the Poincaré section of the transversal (and trapping region) Q_0 under the system (12) is a GAH with an image that is simply a symmetric reflection about the vertical axis of the horseshoe in Figure 1. However, it appears that the construction of an electronic circuit simulating (12) would be a rather formidable undertaking, so we selected a simpler system, namely, the Rössler attractor model, which is a mildly nonlinear three-dimensional ODE that has a straightforward circuit realization.

4. Poincaré Maps and Circuit Realization of the Rössler Attractor

We consider the Rössler attractor

$$\begin{aligned} \dot{x} &= y - z \\ \dot{y} &= x + ay \\ \dot{z} &= b + z(x - c); \end{aligned} \tag{14}$$

where we use the parameters $a = 0.2$, $b = 0.1$, and $c = 10$. This produces the chaotic strange attractor in Figure 3, and it can also be realized by a rather simple electronic circuit.

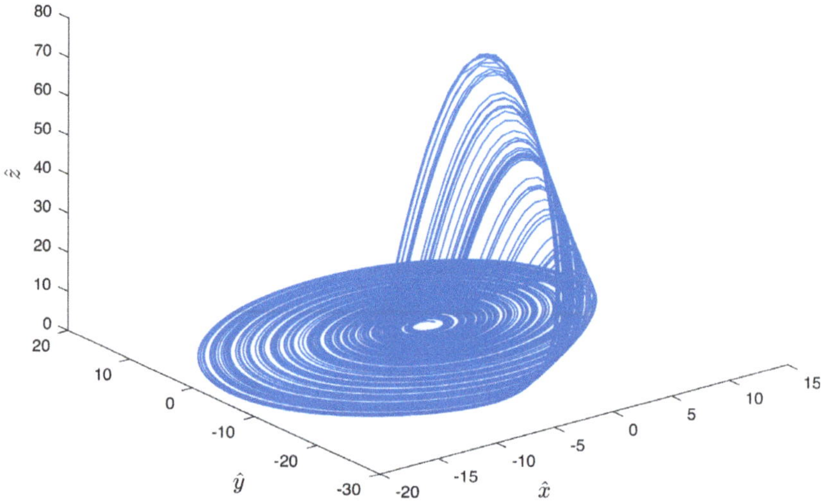

Figure 3. The Rössler attractor with parameters $a = 0.2$, $b = 0.1$, and $c = 10$, and a rotation (represented by $\hat{x}$ and $\hat{y}$) of $\theta = 2\pi/5$ in spherical coordinates.

4.1. The Poincaré Map

One can use the algorithm in Section 2 to compute any Poincaré section of the attractor; however, what we are particularly interested in is identifying a trapping region for a generalized attracting horseshoe. Assuming the system contains a GAH, we first look for a Poincaré section with a horseshoe-like structure as shown in Figure 4.

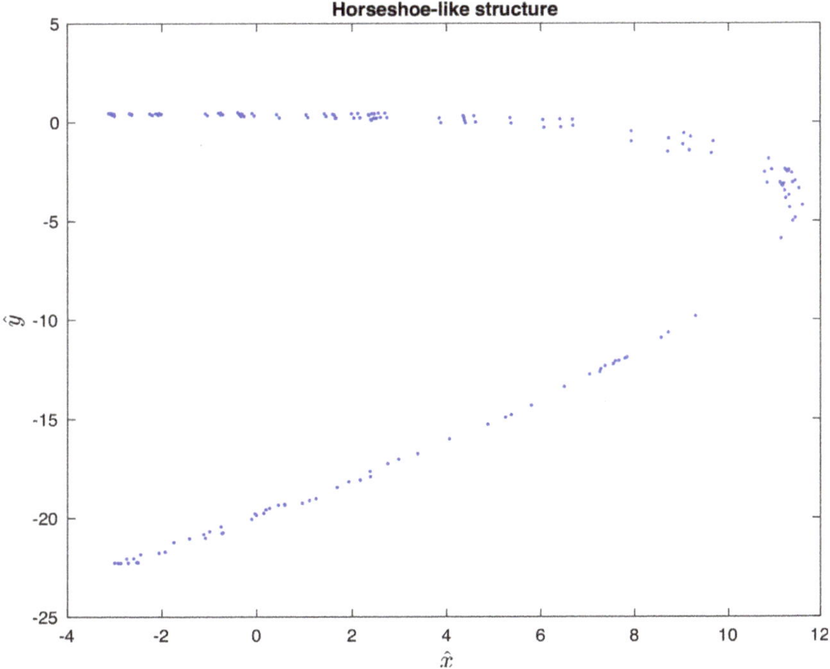

Figure 4. Poincaré section ($r = 5, \theta = 2\pi/5$) of the Rössler attractor containing a horseshoe-like structure. Plot is shown in the rotated frame.

Now, if we can find a rotationally symmetric trapping region around this horseshoe, we shall have shown evidence for the existence of a GAH. First, we identify vertices of a quadrilateral that fully encompasses the horseshoe-like structure. Then, using a recursive algorithm (described in Section 2) we compute the first return map of those vertices on that particular Poincaré section, i.e., the first iteration of the Poincaré map of those points. If the iterates are contained within that quadrilateral, the points on the quadrilateral itself can be tested. In Figure 5, four-thousand points on the quadrilateral are iterated, and it is illustrated that this first return is completely contained in the quadrilateral. While this is not a proof, the grid spacing on the quadrilateral provides compelling evidence that this is a trapping region for the GAH.

In order to provide more compelling evidence, we compute higher-order iterations of the Poincaré map in Figure 6.

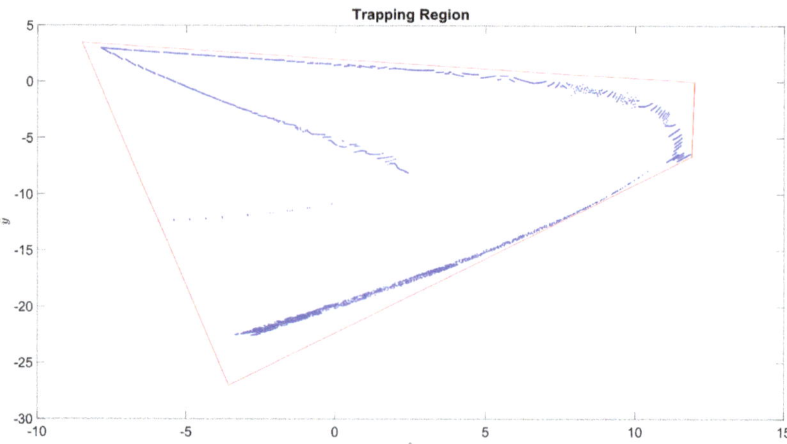

Figure 5. The first return (blue markers) of the quadrilateral trapping region (red markers) with vertices located at $(\hat{x}, \hat{y}) = (-3.55, -27)$, $(11.91, -6.6)$, $(12, 0)$, $(-8.5, 3.5)$. While the quadrilateral edges look "continuous", it should be noted that it is in fact discretized using four thousand points, which are then mapped back to the Poincaré section ($r = 5, \theta = 2\pi/5$). Plot is shown in the rotated frame with $\hat{x}$ and $\hat{y}$ denoting rotated axes.

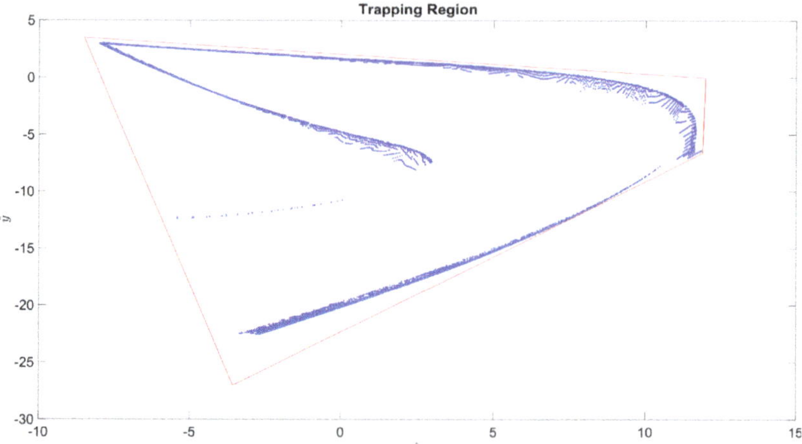

Figure 6. First five iterations of the Poincaré map (blue markers) of the quadrilateral trapping region (red markers) with vertices located at $(\hat{x}, \hat{y}) = (-3.55, -27)$, $(11.91, -6.6)$, $(12, 0)$, $(-8.5, 3.5)$. While the quadrilateral edges look "continuous", it should be noted that it is in fact discretized using four thousand points, which are then mapped back to the Poincaré section ($r = 5, \theta = 2\pi/5$). Plot is shown in the rotated frame with $\hat{x}$ and $\hat{y}$ denoting rotated axes.

4.2. Circuit Realization of the Rössler System

It happens that there are several known examples of electronic circuits realizing the Rössler attractor system. We chose the one, obtained from [11], shown in Figure 7 with a list of components in Table 1.

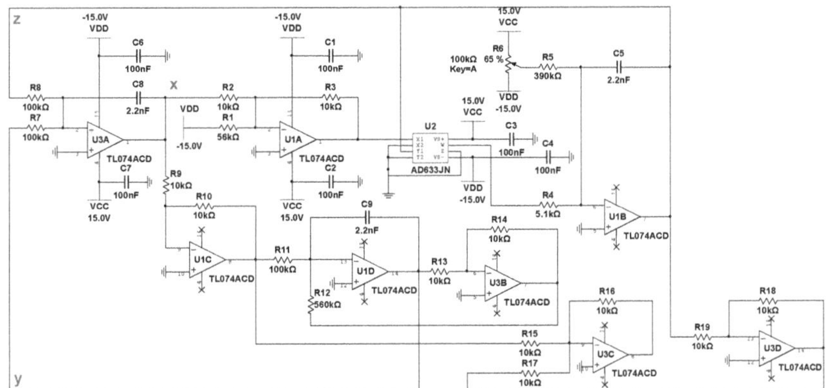

Figure 7. Multisim circuit diagram for Röossler attractor.

The physical realization of the Rössler attractor circuit was constructed using summing amplifiers, integrators, and a multiplier. Due to the nature of this system, the operational amplifier must operate within ±15 volts in order to avoid clipping of the Rössler Attractor output waveform. In this circuit, resistors were used to represent constant values for parameters a and b in (14). A potentiometer was used to vary the parameter value of b in order to observe the bifurcations of the physical system. We first test the circuit on *Multisim* and observe the aforementioned bifurcations in Figure 8.

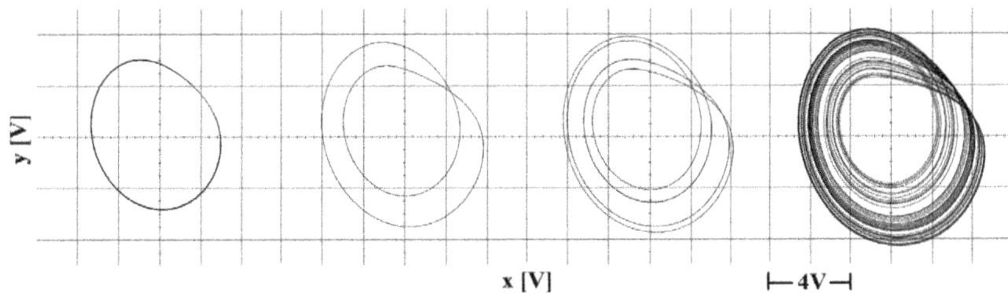

Figure 8. Multisim outputs of the Rössler attractor showing a period doubling Hopf bifurcation leading to chaos.

Table 1. List of components for the Röossler attractor circuit.

Type	Quantity	Code
10 kΩ Resistor	11	
100 kΩ Resistor	3	
390 kΩ Resistor	1	
56 kΩ Resistor	1	
560 kΩ Resistor	1	
5.1 kΩ Resistor	1	
100 kΩ Potentiometer	1	
100 nF Capacitor	6	
2.2 nF Capacitor	3	
Op-Amp	2	AD633JN
Multiplier	1	TL074CN

Next, we built the circuit and observed oscilloscope outputs as shown in Figure 9. The Poincaré section that we chose was a particular vertical plane through the top arch of the output shown (see also Figure 3). The acceptable planes were obtained by trial and error via varying the system parameters and rotation of the plane about a vertical axis through the apex of the arc.

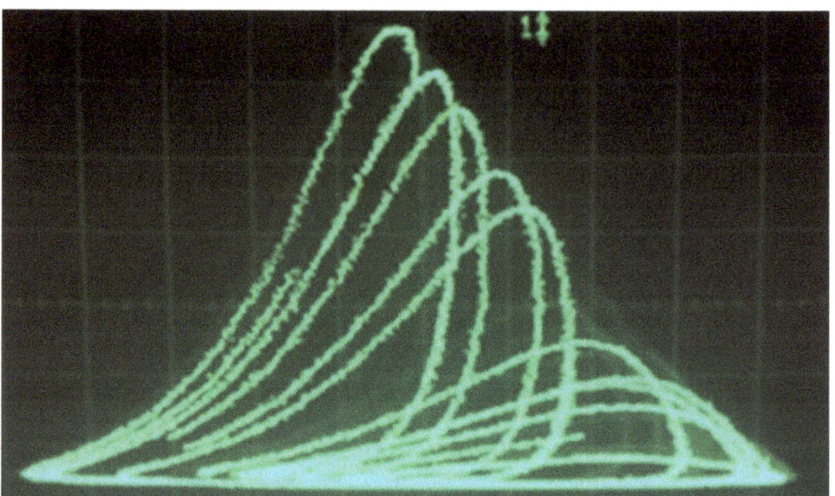

Figure 9. Oscilloscope output from Rössler attractor circuit.

5. Potential Applications

One can imagine several practical applications of devices containing electronic circuit realizations of a GAH. Two, which are related to communications and intelligence gathering, immediately come to mind: First, the circuit could be embedded in a communication receiving device, and tuned to certain "static" frequencies different from those in the expected incoming messages. The strong global attracting characteristics of the circuit would separate the static from the incoming messages, thereby enhancing the receiving capabilities of system. In effect, the GAH circuit would filter out the static.

Second, a stationary or compact mobile device incorporating the GAH circuit could be used to penetrate and analyze various communication systems. Either by connecting remotely in the case of a stationary device or directly for a mobile version, the global attracting properties could be employed to extract crucial characteristics of the system to which it is connected. Moreover, the same attracting features of the GAH circuit device could be used to absorb various parts of sent messages that would render them useless, false, or simply misleading.

The two rather basic applications mentioned provide just a glimpse of the possible applications of GAH circuits, most of which would probably be related to information systems, data collection, and filtering. Moreover, there are more applications that could exploit the chaotic strange attractor associated with a GAH circuit. For, example a GAH circuit device could be used either to control chaos, introduce chaos or adjust the fractal dimension of outputs of a variety of applicable processes based on dynamical systems.

Such mechanisms may aid in a variety of fields including cryptography and cyber security. While encryption techniques reliant on the iterate by iterate behavior of chaotic maps have had their short comings due to irreversibly in analog form and a lack of proper security when implemented in software, the global properties of a GAH circuit may provide a useful intermediate stage during various forms of symmetric encryption by allowing a signal of importance to pass both through and around such a subsystem in parallel. Long-term global properties from the signal sent through a GAH circuit can be extracted and used to manipulate the unaltered signal before reaching the recipient, thereby substantially increasing the difficulty of decryption. As the global properties of such a system (i.e., factors taken from its geometry and macro-scale structure) are less influenced by minor imperfections in the circuit, reversing the process becomes far more tangible.

Furthermore, adding to our second practical application claim, a GAH device could prove to have various applications in machine learning, specifically regarding the creation and prevention of adversarial attacks on deep neural networks. By altering carefully chosen properties of the data meant to be received by the network intelligible yet incorrect results can be forced. In the same light, the proper extraction of dominant incoming signals may help prevent the same sort of issue in special cases.

6. Conclusions

We constructed a rather complicated nonlinear three-dimensional ordinary differential equation (ODE) having a Poincaré section that is a GAH map, but it is not particularly amenable to electronic circuit realization, which was a goal of the investigation. Therefore, instead of the initial ODE, we selected the Rössler attractor; a mildly nonlinear three-dimensional ODE that has a reasonably simple circuit realization and can actually produce GAH maps for carefully chosen Poincaré sections. We constructed the corresponding GAH circuit and used a novel iteration procedure to generate good approximations of the chaotic strange attractors associated to the GAH maps. Finally, in addition to the experimental and analytic aspects of our investigation, we discussed a number of potential practical applications of the GAH circuit. Most of the envisioned applications were in the realms of communication and information gathering.

Supplementary Materials: The following are available at https://www.mdpi.com/2073-8994/13/1/30/s1.

Author Contributions: Conceptualization, A.R. and D.B.; Formal analysis, K.M.; Investigation, K.M., I.J., P.S. and A.R.; Methodology, A.R.; Project administration, A.R.; Supervision, A.R. and D.B.; Validation, I.J.; Visualization, K.M.; Writing—original draft, A.R. and D.B.; Writing—review & editing, I.J., A.R. and D.B. All authors have read and agreed to the published version of the manuscript.

Funding: This research received no external funding.

Institutional Review Board Statement: Not applicable.

Informed Consent Statement: Not applicable.

Data Availability Statement: All data and computer programs that were developed are included as Supplementary materials.

Acknowledgments: The authors would like to thank the NJIT Provost Research Grants for funding this research. K.M. was funded by the Provost high school internship, and P.S. was funded by Phase-1 Provost Undergraduate Research Grant, with D.B. as faculty mentor and A.R. as graduate student mentor. K.M. appreciates the support of Bridgewater-Raritan High School, P.S. and I.J. appreciate the support of the Electrical and Computer Engineering Department at NJIT, and A.R. and D.B. appreciate the support of the Department of Mathematical Sciences and the Center for Applied Mathematics and Statistics (CAMS) at NJIT. In addition, K.M., I.J., and A.R. would like to acknowledge their current institutions: K.M. appreciates the support of the Department of Computer Science at UI-UC, I.J. appreciates the support of the Department of Applied Mathematics and Statistics at SU, and A.R. appreciates the support of the Department of Applied Mathematics at UW. Finally, the authors would like to give their sincere thanks to the reviewers for their insightful comments that helped improve the manuscript.

Conflicts of Interest: The authors declare no conflict of interest.

References

1. Smale, S. Diffeomorphisms with many periodic points. In *Differential and Combinatorial Topology*; Carins, S., Ed.; Princeton University Press: Princeton, NJ, USA, 1963; pp. 63–80.
2. Lorenz, E.N. Deterministic nonperiodic flow. *J. Atoms. Sci.* **1963**, *20*, 130–141. [CrossRef]
3. Rössler, O.E. An equation for continuous chaos. *Phys. Lett. A* **1976**, *57*, 397–398. [CrossRef]
4. Matsumoto, T. A chaotic attractor from Chua's circuit. *IEEE Trans. Circuits Syst.* **1984**, *31*, 1055–1058. [CrossRef]
5. Hénon, M. A two-dimensional mapping with a strange attractor. *Commun. Math. Phys.* **1976**, *50*, 69–77. [CrossRef]

6. Lozi, R. Un attracteur étrange (?) du type attracteur de hénon. *J. Phys. Colloq.* **1978**, *39*, 9–10. [CrossRef]
7. Blackmore, D.; Joshi, Y. Strange attractors for asymptotically zero maps. *Chaos Solitons Fractals* **2014**, *68*, 123–138.
8. Joshi, Y.; Blackmore, D.; Rahman, A. Generalized attracting horseshoe and chaotic strange attractors. *arXiv* **2020**, arXiv:1611.04133v2.
9. Rahman, A.; Jordan, I.; Blackmore, D. Qualitative models and experimental investigation of chaotic nor gates and set/reset flip-flops. *Proc. Roy. Soc. A* **2018**, *474*, 1–19. [CrossRef] [PubMed]
10. Gonze, D. poincare.m. Available online: http://homepages.ulb.ac.be/~dgonze/info/matlab/poincare.m (accessed on 25 February 2011).
11. Glen, K. Rossler Attractor. Available online: http://www.glensstuff.com/rosslerattractor/rossler.htm (accessed on 25 January 2017).

Article

Multiple Solutions for a Class of Nonlinear Fourth-Order Boundary Value Problems

Longfei Lin , Yansheng Liu * and Daliang Zhao

School of Mathematics and Statistics, Shandong Normal University, Jinan 250014, China; 2019020438@stu.sdnu.edu.cn (L.F.); dlzhao928@sdnu.edu.cn (D.Z.)

* Correspondence: ysliu@sdnu.edu.cn

Received: 14 November 2020; Accepted: 29 November 2020; Published: 2 December 2020

Abstract: This paper is concerned with multiple solutions for a class of nonlinear fourth-order boundary value problems with parameters. By constructing a special cone and applying fixed point index theory, the multiple solutions for the considered systems are obtained under some suitable assumptions. The main feature of obtained solutions $(u(t),\ v(t))$ is that the solution $u(t)$ is positive, and the other solution $v(t)$ may change sign. Finally, two examples with continuous function f_1 being positive and f_2 being semipositone are worked out to illustrate the main results.

Keywords: multiple solutions; fixed point theory; boundary value problems

---

1. Introduction

It is well known that the subject of the existence of solutions to numerous boundary value problems (BVP) for differential equations such as second-order [1–3], fourth-order [4–6], even fractional order BVP [7–11] has gained considerable attention and popularity. A growing number of outstanding progress has been made in the theory of such BVP in the last decades due mainly to their extensive applications in the fields of hydrodynamics, nuclear physics, biomathematics, chemistry, and control theory. For further details, please see References [12–29] and references therein.

It is noted that fourth-order boundary value problems have an important application in practical problems, that is, they can be used to describe the deformation of elastic beam, see References [30–33] and references therein. For example, in Reference [32], by means of the theory of fixed point index on cone, Y. Li investigated the following boundary value problems of fourth-order ordinary differential equation

$$\begin{cases} u^{(4)}(t)+\beta u''(t)-\alpha u(t)=f(t,u), \ \ 0<t<1; \\ u(0)=u(1)=u''(0)=u''(1)=0, \end{cases}$$

where $f\in C([0,1]\times\mathbb{R}^+,\mathbb{R}^+)$, $\alpha,\beta\in\mathbb{R}$ and satisfy $\beta<2\pi^2,\alpha\geq-\beta^4/4,\alpha/\pi^4+\beta/\pi^2<1$. By constructing a special cone, the existence of at least one positive solution was obtained under some suitable assumptions.

Recently, in Reference [33], Q. Wang and L. Yang studied the following boundary value problems

$$\begin{cases} u^{(4)}(t)+\beta_1 u''(t)-\alpha_1 u(t)=f_1(t,u(t),v(t)), \ \ 0<t<1; \\ v^{(4)}(t)+\beta_2 v''(t)-\alpha_2 v(t)=f_2(t,u(t),v(t)), \ \ 0<t<1; \\ u(0)=u(1)=u''(0)=u''(1)=0; \\ v(0)=v(1)=v''(0)=v''(1)=0, \end{cases} \tag{1}$$

where $f_1, f_2 \in C([0,1] \times \mathbb{R}^+ \times \mathbb{R}^+, \mathbb{R}^+)$, and $\beta_i, \alpha_i \in \mathbb{R}(i=1,2)$ satisfy the following conditions:

$$\beta_i < 2\pi^2, \ -\beta_i/4 \le \alpha_i, \ \alpha_i/\pi^4 + \beta_i/\pi^2 < 1. \tag{2}$$

These conditions involve a two-parameter non-resonance condition. By constructing two classes of cones and using the fixed point theory, the existence of at least one positive solution was obtained. It is remarkable that the premise of this establishment of the result in Reference [33] is that the nonlinear term f_2 must be positive.

We point out that there are some limitations in those existing results of fourth-order boundary value problems. All solutions obtained in the above references are positive, and moreover, the corresponding conclusions in them are not valid when the nonlinear term is allowed to be non-positive. Considering that two variables u and v in the nonlinear term usually have some connections in many practical problems, there is no description of the relationship between them in the aforementioned papers. It is an interesting problem to seek such solutions for BVP (1) that one variable is positive and the other may be non-positive under the assumptions that nonlinearity may be semipositoned, and some connection will be added between these two variables. As far as we know, there is no paper considering such problem for BVP (1). The purpose of the present paper is to fill this gap.

This paper, motivated by all the above mentioned discussions, investigates the multiple solutions for BVP (1) under the more different conditions compared with Reference [33]. By constructing a very special cone and using the fixed point index theory, the existence and multiplicity results of solutions to (1) are obtained when $\beta_i, \alpha_i \in \mathbb{R}$ $(i=1,2)$ satisfy the conditions (2), $f_1 \in C([0,1] \times \mathbb{R}^+ \times \mathbb{R}, \mathbb{R}^+)$, and $f_2 \in C([0,1] \times \mathbb{R}^+ \times \mathbb{R}, \mathbb{R})$.

The nonlinear term f_2 is allowed to change sign by contrast, $f_2 \in C([0,1] \times \mathbb{R}^+ \times \mathbb{R}, \mathbb{R})$. A relationship is imposed between two variables u, v in nonlinear terms, which is that the variable v is controlled by u. In obtained solution (u,v), the component u is positive, but the component v is allowed to be negative in comparison with Reference [33].

The rest of this paper is organized as follows—Section 2 contains some background materials and preliminaries. The main results will be given and proved in Section 3. Finally, in Section 4, two examples are given to support our results.

2. Background Materials and Preliminaries

The basic space used in this paper is $E := C[0,1] \times C[0,1]$. It is a Banach space endowed with the norm $\|(u,v)\| = \max\{\|u\|, \|v\|\}$ for $(u,v) \in E$, where $\|u\| = \max_{t\in[0,1]} |u(t)|$, $\|v\| = \max_{t\in[0,1]} |v(t)|$. Under the condition (2), as in Reference [32], let

$$\xi_{i,1} = \frac{-\beta_i + \sqrt{\beta_i^2 + 4\alpha_i}}{2}, \ \xi_{i,2} = \frac{-\beta_i - \sqrt{\beta_i^2 + 4\alpha_i}}{2}, \ (i=1,2),$$

and let $G_{i,j}(t,s)(i,j=1,2)$ be the Green's function of the linear boundary value problem

$$\begin{cases} -u_i''(t) + \xi_{i,j} u_i(t) = 0, \ 0 < t < 1; \\ u_i(0) = u_i(1) = 0, \ i,j = 1,2. \end{cases}$$

Then for $h_i \in C[0,1]$, the solution of the following nonlinear boundary value problem

$$\begin{cases} u_i^{(4)}(t) + \beta_i u_i''(t) - \alpha_i u_i = h_i(t), \ 0 < t < 1; \\ u_i(0) = u_i(1) = u_i''(0) = u_i''(1) = 0, \ i,j = 1,2 \end{cases}$$

can be expressed as

$$u_i(t) = \int_0^1 \int_0^1 G_{i,1}(t,\tau)G_{i,2}(\tau,s)h_i(s)dsd\tau, \ \ t \in [0,1].$$

Lemma 1. *The function $G_{i,j}(t,s)(i = 1,2)$ has the following properties:*
(1) $G_{i,j}(t,s) > 0$ for $t,s \in (0,1)$;
(2) $G_{i,j}(t,s) \le C_{i,j}G_{i,j}(s,s)$ for $t,s \in [0,1]$, where $C_{i,j} > 0$ is a constant;
(3) $G_{i,j}(t,s) \ge \delta_{i,j}G_{i,j}(t,t)G_{i,j}(s,s)$ for $t,s \in [0,1]$, where $\delta_{i,j} > 0$ is a constant;
(4) $G_{2,j}(t,s) \le N_j G_{1,j}(t,s)$ for $t,s \in [0,1]$, where $N_j > 0$ is a constant.

Proof of Lemma 1. (1)–(3) can be seen from Reference [32]. In addition, by careful calculation and Lemma 2.1 in Reference [32], it is not difficult to prove that $N_j := \sup\limits_{0<t,s<1} \frac{G_{2,j}(t,s)}{G_{1,j}(t,s)} < +\infty$. Immediately, (4) is derived. □

The main tool used here is the following fixed-point index theory.

Lemma 2 ([34])**.** *Let E_1 be a Banach space and P be a cone in E_1. Denote $P_r = \{u \in P : \|u\| < r\}$ and $\partial P_r = \{u \in P : \|u\| = r\}$ $(\forall r > 0)$. Let $T : P \to P$ be a complete continuous mapping, then the following conclusions are valid.*

(1) If $\mu Tu \ne u$ for $u \in \partial P_r$ and $\mu \in (0,1]$, then $i(T, P_r, P) = 1$;

(2) If $\inf\limits_{u\in\partial P_r} \|Tu\| > 0$ and $\mu Tu \ne u$ for $u \in \partial P_r$ and $\mu \ge 1$, then $i(T, P_r, P) = 0$.

3. Main Results

In this section, we shall establish the existence and multiplicity results, which is based on the fixed point index theory. For this matter, first we define the mappings $T_1, T_2 : E \to C[0,1]$, and $T : E \to E$ by

$$T_1(u,v)(t) = \int_0^1 \int_0^1 G_{1,1}(t,\tau)G_{1,2}(\tau,s)f_1(s,u(s),v(s))dsd\tau,$$

$$T_2(u,v)(t) = \int_0^1 \int_0^1 G_{2,1}(t,\tau)G_{2,2}(\tau,s)f_2(s,u(s),v(s))dsd\tau,$$

$$T(u,v)(t) = (T_1(u,v)(t), T_2(u,v)(t)), \ \ \forall (u,v) \in E.$$

Then, BVP (1) in operator forms becomes

$$(u,v) = T(u,v). \tag{3}$$

By (3), one can easily see that the existence of solutions for BVP (1) is equivalent to the existence of nontrivial fixed point of T. Therefore, we need to find only the nontrivial fixed point of T in the following work.

Subsequently, for simplicity and convenience, set

$$M_{i,j} = \max_{t\in[0,1]} G_{i,j}(t,t), \quad C_i = \int_0^1 G_{i,1}(\tau,\tau)G_{i,2}(\tau,\tau)d\tau, \quad \text{and } \lambda_i = \pi^4 - \beta_i\pi^2 - \alpha_i.$$

Then, $M_{i,j}, C_i$, and $\lambda_i (i,j = 1,2)$ are positive numbers.

Now let us list the following assumptions satisfied throughout the paper.

(H1) $f_1 \in C([0,1] \times \mathbb{R}^+ \times \mathbb{R}, \mathbb{R}^+)$, $f_2 \in C([0,1] \times \mathbb{R}^+ \times \mathbb{R}, \mathbb{R})$, and there exists $N_3 > 0$ such that $|f_2(t,u,v)| \le N_3 f_1(t,u,v)$ for $(t,u,v) \in [0,1] \times \mathbb{R}^+ \times \mathbb{R}$.

(H2) $\lim\limits_{\substack{|v|\le Nu\\ u\to 0^+}} \sup \max\limits_{t\in[0,1]} \frac{f_1(t,u,v)}{u} < \lambda_1 < \lim\limits_{\substack{|v|\le Nu\\ u\to +\infty}} \inf \min\limits_{t\in[0,1]} \frac{f_1(t,u,v)}{u}$.

(H3) $\lim\limits_{\substack{|v|\le Nu\\ u\to 0^+}} \inf \min\limits_{t\in[0,1]} \frac{f_1(t,u,v)}{u} > \lambda_1 > \lim\limits_{\substack{|v|\le Nu\\ u\to +\infty}} \sup \max\limits_{t\in[0,1]} \frac{f_1(t,u,v)}{u}$.

In addition, for the sake of obtaining the nontrivial fixed point of operator T, let

$$P = \{(u,v) \in E : u(t) \ge \sigma(t)\|u\| \text{ and } |v(t)| \le Nu(t), \forall t \in [0,1]\},$$

where $\sigma(t) = \frac{\delta_{1,1}\delta_{1,2}C_1}{C_{1,1}C_{1,2}M_{1,1}} G_{1,1}(t,t)$ and $N = N_1N_2N_3$. N_1, N_2, and N_3 are defined in Lemma 1 and (H1), respectively.

Obviously, P is a nonempty, convex, and closed subset of E. Furthermore, one can prove that P is a cone of Banach space E.

For convenience, set

$$\Lambda_Y = \{(u,v) \in \mathbb{R}^+ \times \mathbb{R} : u \in Y \subset \mathbb{R}^+, |v| \le Nu\},$$

$$\begin{aligned} P_r &= \{(u,v) \in P : \|u\| < r\}, \\ \partial P_r &= \{(u,v) \in P : \|u\| = r\}, \\ \bar{P}_r &= \{(u,v) \in P : \|u\| \le r\}. \end{aligned}$$

It is not difficult to see that P_r is a relatively open and bounded set of P for each $r > 0$.

Lemma 3. *To calculate the fixed point index of T in P_r, we first need to prove the following result. Assume that (H_1) hold. Then $T : P \to P$ is completely continuous, and $T(P) \subset P$.*

Proof of Lemma 3. For $(u,v) \in P$, by virtue of Lemma 1, one can easily obtain that

$$\begin{aligned} T_1(u,v)(t) &= \int_0^1 \int_0^1 G_{1,1}(t,\tau)G_{1,2}(\tau,s)f_1(s,u(s),v(s))dsd\tau \\ &\ge \frac{\delta_{1,1}\delta_{1,2}C_1}{C_{1,1}C_{1,2}M_{1,1}} G_{1,1}(t,t)\|T_1(u,v)\| = \sigma(t)\|T_1(u,v)\|, \quad \forall t \in [0,1]. \end{aligned}$$

Moreover, (H1) together with Lemma 1 guarantees that

$$\begin{aligned} |T_2(u,v)(t)| &= |\int_0^1 \int_0^1 G_{2,1}(t,\tau)G_{2,2}(\tau,s)f_2(s,u(s),v(s))dsd\tau| \\ &\le N_3 \int_0^1 \int_0^1 G_{2,1}(t,\tau)G_{2,2}(\tau,s)f_1(s,u(s),v(s))dsd\tau \\ &\le N_1N_2N_3 \int_0^1 \int_0^1 G_{1,1}(t,\tau)G_{1,2}(\tau,s)f_1(s,u(s),v(s))dsd\tau \\ &= N|T_1(u,v)(t)|. \end{aligned}$$

Therefore, $T(u,v) \in P$, namely, $T(P) \subset P$. In addition, since f_1, f_2, and $G_{i,j}$ are continuous, one can deduce that T is completely continuous by using normal methods such as Arscoli-Arzela theorem, and so forth. □

Now we are in a position to prove our main results in the following.

Theorem 1. *Under the assumptions (H1) and (H2), the BVP (1) admits at least one nontrivial solution.*

Proof of Theorem 1. To obtain the nontrivial solution for BVP (1), we will choose a bounded open set $P_{R_1} \setminus \bar{P}_{r_1}$ in cone P and calculate the fixed point index $i(T, P_{R_1} \setminus \overline{P_{r_1}}, P)$. For this, the proof of Theorem 1 will be carried out in three steps.

First, notice that by (H2), there exist $\varepsilon \in (0,1)$ and $r_1 > 0$ such that

$$f_1(t,u,v) \leq (1-\varepsilon)\lambda_1 u \quad \forall t \in [0,1], (u,v) \in \Lambda_{[0,r_1]}. \tag{4}$$

We claim that

$$\mu T(u,v) \neq (u,v),\ \forall \mu \in (0,1],\ (u,v) \in \partial P_{r_1}. \tag{5}$$

To this end, suppose on the contrary that there exist $\mu_0 \in (0,1]$ and $(u_0, v_0) \in \partial P_{r_1}$ such that

$$\mu_0 T(u_0, v_0) = (u_0, v_0).$$

Therefore, (u_0, v_0) satisfies the following differential equation

$$\begin{cases} u_0^{(4)}(t) + \beta_1 u_0''(t) - \alpha_1 u_0(t) = f_1(t, u(t), v(t)), \quad 0 < t < 1; \\ u_0(0) = u_0(1) = u_0''(0) = u_0''(1) = 0; \end{cases} \tag{6}$$

It follows from (4) and (6) that

$$u_0^{(4)}(t) + \beta_1 u_0''(t) - \alpha_1 u_0(t) \leq f_1(t, u_0(t), v_0(t)) \leq (1-\varepsilon)\lambda_1 u_0(t).$$

Multiplying the above inequality by $\sin(\pi t)$ and then integrating from 0 to 1, one can easily get

$$\int_0^1 \lambda_1 u_0(t) \sin(\pi t) dt \leq (1-\varepsilon) \int_0^1 \lambda_1 u_0(t) \sin(\pi t) dt.$$

Noticing that $\int_0^1 \lambda_1 u_0(t) \sin(\pi t) dt > 0$, we obtain a contradiction.

Second, from (H2), there exist $\varepsilon > 0$ and $m > 0$ such that

$$f_1(t,u,v) \geq (1+\varepsilon)\lambda_1 u \quad \forall t \in [0,1], (u,v) \in \Lambda_{[m,+\infty)}. \tag{7}$$

Set $C := \max\limits_{\substack{t \in [0,1] \\ (u,v) \in \Lambda_{[0,m]}}} |f_1(t,u,v) - (1+\varepsilon)\lambda_1 u| + 1$. Then one can easily find that

$$f_1(t,u,v) \geq (1+\varepsilon)\lambda_1 u - C,\ \forall t \in [0,1],\ (u,v) \in \Lambda_{\mathbb{R}^+}. \tag{8}$$

Now, we will show that there exists $R_1 > r_1$ such that

$$\inf_{(u,v) \in \partial P_{R_1}} \|T(u,v)\| > 0 \ \text{ and } \ \mu T(u,v) \neq (u,v),\ \forall \mu \geq 1,\ (u,v) \in \partial P_{R_1}. \tag{9}$$

Suppose, on the contrary, that there exist $\mu_0 \geq 1$ and $(u_0, v_0) \in \partial P_{R_1}$ such that $\mu_0 T(u_0, v_0) = (u_0, v_0)$. Combining (6) with (8), we immediately get

$$u_0^{(4)}(t) + \beta_1 u_0''(t) - \alpha_1 u_0(t) \geq f_1(t, u_0(t), v_0(t)) \geq (1+\varepsilon)\lambda_1 u_0(t) - C.$$

Hence,

$$\int_0^1 \lambda_1 u_0(t) \sin(\pi t) dt \geq (1+\varepsilon) \int_0^1 \lambda_1 u_0(t) \sin(\pi t) dt - \frac{2C}{\pi},$$

which yields

$$\int_0^1 \lambda_1 u_0(t) \sin(\pi t) dt \leq \frac{2C}{\pi \varepsilon \lambda_1}.$$

On the other hand, in view of the definition of cone P, one can easily obtain that

$$\|u_0\| \int_0^1 \sigma(t)\sin(\pi t)dt \le \int_0^1 u_0(t)\sin(\pi t)dt \le \frac{2C}{\pi\varepsilon\lambda_1},$$

which means

$$\|u_0\| \le \frac{2C}{\pi\varepsilon\lambda_1 \int_0^1 \sigma(t)\sin(\pi t)dt} := R_1^*. \tag{10}$$

Therefore, if $R_1 > R_1^*$, immediately, one can get $\mu T(u,v) \neq (u,v)$ for $\mu \ge 1$ and $(u,v) \in \partial P_{R_1}$. In addition, if $R_1 > \dfrac{m}{\min\limits_{t\in[\frac{1}{4},\frac{3}{4}]} \sigma(t)} := \dfrac{m}{\sigma^*}$, then by the definition of cone P, one can get that for any $t \in [\frac{1}{4}, \frac{3}{4}]$ and $(u,v) \in \partial P_{R_1}$,

$$u(t) \ge \min_{t\in[\frac{1}{4},\frac{3}{4}]} u(t) \ge \sigma^* R_1 > m. \tag{11}$$

So, by (7), (11), and Lemma 1, one can get that for all $(u,v) \in \partial P_{R_1}$,

$$\begin{aligned}
\|T(u,v)\| &\ge T_1(u,v)(\frac{1}{2}) \\
&= \int_0^1 \int_0^1 G_{1,1}(\frac{1}{2},\tau)G_{1,2}(\tau,s)f_1(s,u(s),v(s))dsd\tau \\
&\ge \delta_{1,1}\delta_{1,2}G_{1,2}(\frac{1}{2},\frac{1}{2})C_1 \int_0^1 G_{1,2}(s,s)f_1(s,u(s),v(s))ds \\
&\ge \delta_{1,1}\delta_{1,2}G_{1,2}(\frac{1}{2},\frac{1}{2})C_1(1+\varepsilon)\lambda_1 \int_{\frac{1}{4}}^{\frac{3}{4}} G_{1,2}(s,s)u(s)ds \\
&\ge \delta_{1,1}\delta_{1,2}G_{1,2}(\frac{1}{2},\frac{1}{2})C_1(1+\varepsilon)\lambda_1 R_1\sigma^* \int_{\frac{1}{4}}^{\frac{3}{4}} G_{1,2}(s,s)ds \\
&\ge \delta_{1,1}\delta_{1,2}G_{1,2}(\frac{1}{2},\frac{1}{2})C_1(1+\varepsilon)\lambda_1 m \int_{\frac{1}{4}}^{\frac{3}{4}} G_{1,2}(s,s)ds > 0.
\end{aligned}$$

That is, $\inf\limits_{(u,v)\in\partial P_{R_1}} \|T(u,v)\| > 0$. So, we can ultimately choose $R_1 > \max\{R_1^*, r_1, \frac{m}{\sigma^*}\}$ such that (9) holds.

Based on (5), (9), Lemma 2, and Lemma 3, we have

$$i(T, P_{R_1} \setminus \bar{P_{r_1}}, P) = i(T, P_{R_1}, P) - i(T, P_{r_1}, P) = 0 - 1 = -1.$$

As a result, the conclusion of this theorem follows. □

Theorem 2. *Assume that (H1) and (H3) hold. Then the BVP (1) has at least one nontrivial solution.*

Proof of Theorem 2. In the following, we divide the proof of Theorem 2 into three steps.

Step 1. From condition (H3), there exist $\varepsilon > 0$ and $r_2 > 0$ such that

$$f_1(t,u,v) \ge (1+\varepsilon)\lambda_1 u, \ \forall t \in [0,1], \ (u,v) \in \Lambda_{[0,r_2]}. \tag{12}$$

Subsequently, we claim that

$$\inf_{(u,v)\in\partial P_{r_2}} \|T(u,v)\| > 0 \ \text{and} \ \mu T(u,v) \neq (u,v), \ \forall \mu \ge 1, \ (u,v) \in \partial P_{r_2}. \tag{13}$$

In fact, if there exist $\mu_0 \geq 1$ and $(u_0, v_0) \in \partial P_{r_2}$ such that $\mu_0 T(u_0, v_0) = (u_0, v_0)$, then by (6) and (12), one can obtain immediately

$$u_0^{(4)}(t) + \beta_1 u_0''(t) - \alpha_1 u_0(t) \geq f_1(t, u_0(t), v_0(t)) \geq (1+\varepsilon)\lambda_1 u_0(t).$$

Hence,

$$\int_0^1 \lambda_1 u_0(t)\sin(\pi t)dt \geq (1+\varepsilon)\int_0^1 \lambda_1 u_0(t)\sin(\pi t)dt.$$

Noticing that $\int_0^1 \lambda_1 u_0(t)\sin(\pi t)dt > 0$, we get a contradiction.

In addition, it follows from Lemma 1 and (12) that for $(u, v) \in \partial P_{r_2}$,

$$\begin{aligned}
\|T(u,v)\| &\geq T_1(u,v)(\frac{1}{2}) \\
&= \int_0^1\int_0^1 G_{1,1}(\frac{1}{2},\tau)G_{1,2}(\tau,s)f_1(s,u(s),v(s))dsd\tau \\
&\geq \delta_{1,1}\delta_{1,2}G_{1,2}(\frac{1}{2},\frac{1}{2})C_1\int_0^1 G_{1,2}(s,s)f_1(s,u(s),v(s))ds \\
&\geq \delta_{1,1}\delta_{1,2}G_{1,2}(\frac{1}{2},\frac{1}{2})C_1(1+\varepsilon)\lambda_1\int_0^1 G_{1,2}(s,s)u(s)ds \\
&\geq \delta_{1,1}\delta_{1,2}G_{1,2}(\frac{1}{2},\frac{1}{2})C_1(1+\varepsilon)\lambda_1 r_2\int_0^1 G_{1,2}(s,s)\sigma(s)ds > 0,
\end{aligned}$$

which yields $\inf\limits_{(u,v)\in\partial P_{r_2}} \|T(u,v)\| > 0$.

Step 2. The assumption (H3) implies that there exist $\varepsilon \in (0,1)$ and $m > 0$ such that

$$f_1(t,u,v) \leq (1-\varepsilon)\lambda_1 u, \ \forall t \in [0,1], \ (u,v) \in \Lambda_{[m,+\infty)}. \tag{14}$$

Moreover, by the continuity of f_1 and f_2, there exists $C^* > 0$ such that

$$f_1(t,u,v) \leq (1-\varepsilon)\lambda_1 u + C^*, \ \forall t \in [0,1], \ (u,v) \in \Lambda_{\mathbb{R}^+}. \tag{15}$$

We claim that there exists a large enough $R_2 > r_2$ such that

$$\mu T(u,v) \neq (u,v), \ \forall \mu \in (0,1], \ (u,v) \in \partial P_{R_2}. \tag{16}$$

Suppose, on the contrary, there exist $\mu_0 \in (0,1]$ and $(u_0, v_0) \in \partial P_{R_2}$ such that $\mu_0 T(u_0, v_0) = (u_0, v_0)$. Then (6) together with (15) guarantees

$$u_0^{(4)}(t) + \beta_1 u_0''(t) - \alpha_1 u_0(t) \leq f_1(t, u_0(t), v_0(t)) \leq (1-\varepsilon)\lambda_1 u_0(t) + C^*.$$

Consequently,

$$\int_0^1 \lambda_1 u_0(t)\sin(\pi t)dt \leq (1-\varepsilon)\int_0^1 \lambda_1 u_0(t)\sin(\pi t)dt + \frac{2C^*}{\pi},$$

namely,

$$\int_0^1 u_0(t)\sin(\pi t)dt \leq \frac{2C^*}{\pi\varepsilon\lambda_1}.$$

Moreover, based on the definition of cone P, we can immediately get

$$\|u_0\|\int_0^1 \sigma(t)\sin(\pi t)dt \leq \int_0^1 u_0(t)\sin(\pi t)dt \leq \frac{2C^*}{\pi\varepsilon\lambda_1},$$

which means

$$\|u_0\| \le \frac{2C^*}{\pi\varepsilon\lambda_1 \int_0^1 \sigma(t)\sin(\pi t)dt} := R_2^*. \tag{17}$$

So, one can choose $R_2 > \max\{R_2^*, r_2\}$ such that (16) holds.

Step 3. From (13), (16), Lemma 2, and Lemma 3, we deduce that

$$i(T, P_{R_2} \setminus \bar{P_{r_2}}, P) = i(T, P_{R_2}, P) - i(T, P_{r_2}, P) = 1 - 0 = 1.$$

As a result, BVP(1) has at least one nontrivial solution. □

Up to now, some existence results of BVP(1) have been obtained by applying the fixed point index theory. In the following, the multiple solutions will be considered for BVP (1).

Theorem 3. *Assume that (H1) holds. In addition, suppose that*

(1) $\lim\limits_{\substack{|v|\le Nu\\ u\to 0^+}} \sup \max\limits_{t\in[0,1]} \frac{f_1(t,u,v)}{u} < \lambda_1,\ \lim\limits_{\substack{|v|\le Nu\\ u\to +\infty}} \sup \max\limits_{t\in[0,1]} \frac{f_1(t,u,v)}{u} < \lambda_1;$

(2) There exists $r > 0$ and a continuous nonnegative function Φ_r such that

$$f_1(t,u,v) \ge \Phi_r(t), \forall (t,u,v) \in [0,1] \times (\sigma(t)r, r) \times [-Nr, Nr]$$

and

$$\max_{t\in[0,1]} \int_0^1 \int_0^1 G_{1,1}(t,\tau)G_{1,2}(\tau,s)\Phi_r(s)dsd\tau > r.$$

Then the BVP (1) has at least two nontrivial solutions.

Proof of Theorem 3. In order to obtain this conclusion, we firstly claim that

$$\inf_{(u,v)\in\partial P_r} \|T(u,v)\| > 0 \text{ and } \mu T(u,v) \ne (u,v), \forall\, \mu \ge 1,\ (u,v) \in \partial P_r. \tag{18}$$

Suppose, on the contrary, there exist $\mu_0 \ge 1$ and $(u_0, v_0) \in \partial P_r$ such that $\mu_0 T(u_0, v_0) = (u_0, v_0)$. Then,

$$\begin{aligned} \|u_0\| &\ge \|T(u_0,v_0)\| \ge T_1(u_0,v_0)(t) \\ &= \int_0^1 \int_0^1 G_{1,1}(t,\tau)G_{1,2}(\tau,s)f_1(s,u(s),v(s))dsd\tau \\ &\ge \int_0^1 \int_0^1 G_{1,1}(t,\tau)G_{1,2}(\tau,s)\Phi_r(s)dsd\tau. \end{aligned} \tag{19}$$

Taking the maximum for both sides of the above inequality in $t \in [0,1]$, we get that

$$\|u_0\| \ge \max_{t\in[0,1]} \int_0^1 \int_0^1 G_{1,1}(t,\tau)G_{1,2}(\tau,s)\Phi_r(s)dsd\tau > r. \tag{20}$$

This means $(u_0, v_0)\overline{\in}\ \partial P_r$, which is a contradiction. Moreover, one can easily see that $\inf\limits_{(u,v)\in\partial P_r} \|T(u,v)\| > 0$ holds from (19) and (20).

Next, similar to the process of proving (5) and (16), there exist $r_1 \in (0, r)$ and $R_2 \ge \max\{R_2^*, r_2, r\}$ such that

$$\mu T(u,v) \ne (u,v), \forall \mu \in (0,1],\ \forall (u,v) \in \partial P_{r_1}, \tag{21}$$

$$\mu T(u,v) \ne (u,v),\ \forall \mu \in (0,1],\ \forall (u,v) \in \partial P_{R_2}. \tag{22}$$

Thus, by (18), (21), (22), Lemma 2, and Lemma 3, one can immediately obtain that

$$i(T, P_{R_2} \setminus \bar{P}_r, P) = i(T, P_{R_2}, P) - i(T, P_r, P) = 1 - 0 = 1,$$

$$i(T, P_r \setminus \bar{P_{r_1}}, P) = i(T, P_r, P) - i(T, P_{r_1}, P) = 0 - 1 = -1.$$

Namely, there exist $(u_1, v_1) \in P_r \setminus \bar{P_{r_1}}$ and $(u_2, v_2) \in P_{R_2} \setminus \bar{P_r}$ satisfying $T(u_i, v_i) = (u_i, v_i)(i = 1, 2)$, that is, $(u_i, v_i)(i = 1, 2)$ is the solution of BVP(1).

Finally, we show $(u_1, v_1) \neq (u_2, v_2)$. To see this we need only to prove BVP(1) has no solution on ∂P_r. Suppose on the contrary, there exists $(u^*, v^*) \in \partial P_r$ being a solution of BVP(1). Then $T(u^*, v^*) = (u^*, v^*)$. By a similar process of obtaining (20), one can get $\|u^*\| = \|T_1(u^*, v^*)\| > r$, which is a contradiction. To sum up, Theorem 3 is proved. □

From a process similar to the above, the following conclusion can be obtained.

Theorem 4. *Suppose that (H1) holds. In addition, suppose that*

(1) $\lim_{\substack{|v| \le Nu \\ u \to 0^+}} \inf \min_{t \in [0,1]} \frac{f_1(t,u,v)}{u} > \lambda_1, \ \lim_{\substack{|v| \le Nu \\ u \to +\infty}} \inf \min_{t \in [0,1]} \frac{f_1(t,u,v)}{\lambda_1 u} > \lambda_1;$

(2) There exists $R > 0$, and a continuous nonnegative function Ψ_R such that

$$f_1(t, u, v) \le \Psi_R(t), \forall (t, u, v) \in [0, 1] \times [\sigma(t)R, R] \times [-NR, NR]$$

and

$$\max_{t \in [0,1]} \int_0^1 \int_0^1 G_{1,1}(t, \tau) G_{1,2}(\tau, s) \Psi_R(s) ds d\tau < R.$$

Then the BVP (1) has at least two nontrivial solutions.

Proof of Theorem 4. We firstly prove that

$$\mu T(u, v) \neq (u, v), \ \forall \mu \in (0, 1], \ (u, v) \in \partial P_R. \tag{23}$$

To this end, suppose on the contrary that there exist $\mu_0 \in (0, 1]$ and $(u_0, v_0) \in \partial P_R$ such that $\mu_0 T(u_0, v_0) = (u_0, v_0)$. Hence, we get $u_0 = \mu_0 T_1(u_0, v_0)$, that is

$$u_0(t) \le T_1(u_0, v_0)(t) \quad \le \int_0^1 \int_0^1 G_{1,1}(t, \tau) G_{1,2}(\tau, s) \Psi_R(s) ds d\tau < R. \tag{24}$$

Noticing that $(u_0, v_0) \in \partial P_R$, this is a contradiction.

Next, from a process similar to (9) and (13), there exist $R_1 > \max\{R, R_1^*, r_1, \frac{m}{\sigma^*}\}$ and $r_2 \in (0, R)$ such that

$$\inf_{(u,v) \in \partial P_{R_1}} \|T(u, v)\| > 0 \ \text{and} \ \mu T(u, v) \neq (u, v), \ \forall \mu \ge 1, \ (u, v) \in \partial P_{R_1}, \tag{25}$$

$$\inf_{(u,v) \in \partial P_{r_2}} \|T(u, v)\| > 0 \ \text{and} \ \mu T(u, v) \neq (u, v), \ \forall \mu \ge 1, \ (u, v) \in \partial P_{r_2}. \tag{26}$$

So, by (23)–(26), Lemma 2, and Lemma 3, one can get

$$i(T, P_{R_1} \setminus \bar{P_R}, P) = i(T, P_{R_1}, P) - i(T, P_R, P) = 0 - 1 = -1,$$

$$i(T, P_R \setminus \bar{P_{r_2}}, P) = i(T, P_R, P) - i(T, P_{r_2}, P) = 1 - 0 = 1.$$

Finally, from a process similar to the end of proof of Theorem 3, BVP(1) has at least two nontrivial solutions. As a result, the conclusion of this theorem follows. □

4. Examples

In this section, two illustrative examples are worked out to show the effectiveness of the obtained results.

Example 1. *Consider the following BVP of fourth-order ordinary differential systems*

$$\begin{cases} u^{(4)}(t) + u''(t) - \pi^2 u(t) = f_1(t,u,v), \ 0 < t < 1; \\ v^{(4)}(t) + \frac{1}{2}v''(t) - \frac{\pi^2}{2}v(t) = f_2(t,u,v), \ 0 < t < 1; \\ u(0) = u(1) = u''(0) = u''(1) = 0; \\ v(0) = v(1) = v''(0) = v''(1) = 0, \end{cases} \tag{27}$$

where

$$f_1(t,u,v) = \begin{cases} (\pi^4 - 2\pi^2)(1+\sin(\pi t))uv^{\frac{1}{4}} & \text{if } 0 < u < 1, \ |v| < u; \\ (\pi^4 - 2\pi^2)(1+\sin(\pi t))v^{\frac{1}{4}} & \text{if } u = 1, \quad |v| < u; \\ (\pi^4 - 2\pi^2)(1+\sin(\pi t))u^{\frac{1}{4}}v^{\frac{1}{4}} & \text{if } u > 1, \quad |v| < u, \end{cases}$$

$$f_2(t,u,v) = \begin{cases} (\pi^4 - 2\pi^2)(1+\cos(\pi t))uv^{\frac{1}{4}} & \text{if } 0 < u < 1, \ |v| < u; \\ (\pi^4 - 2\pi^2)(1+\cos(\pi t))v^{\frac{1}{4}} & \text{if } u = 1, \quad |v| < u; \\ (\pi^4 - 2\pi^2)(1+\cos(\pi t))u^{\frac{1}{4}}v^{\frac{1}{4}} & \text{if } u > 1, \quad |v| < u, \end{cases}$$

Then, BVP (27) has at least two nontrivial solutions.

Proof of Example 1 . BVP (27) can be regarded as a BVP of the form (1). Choosing $\alpha_1 = \pi^2$, $\beta_1 = 1$, and $\lambda_1 = \pi^4 - 2\pi^2 > 0$, then we have

$$\xi_{1,1} = \frac{-\beta_1 + \sqrt{\beta_1^2 + 4\alpha_1}}{2} = \frac{-1+\sqrt{1+4\pi^2}}{2}, \ \xi_{1,2} = \frac{-\beta_1 - \sqrt{\beta_1^2 + 4\alpha_1}}{2} = \frac{-1-\sqrt{1+4\pi^2}}{2}.$$

Clearly, α_1 and β_1 satisfy the condition (2). Moreover, by careful calculation and Lemma 2.1 in Reference [32], one can obtain that

$$G_{1,1}(t,s) = \begin{cases} \dfrac{\sinh w_{1,1}t \sinh w_{1,1}(1-s)}{w_{1,1}\sinh w_{1,1}} & 0 \le t \le s \le 1; \\ \dfrac{\sinh w_{1,1}s \sinh w_{1,1}(1-t)}{w_{1,1}\sinh w_{1,1}} & 0 \le s \le t \le 1, \end{cases}$$

$$G_{1,2}(t,s) = \begin{cases} \dfrac{\sin w_{1,2}t \sin w_{1,2}(1-s)}{w_{1,2}\sin w_{1,2}} & 0 \le t \le s \le 1; \\ \dfrac{\sin w_{1,2}s \sin w_{1,2}(1-t)}{w_{1,2}\sin w_{1,2}} & 0 \le s \le t \le 1, \end{cases}$$

where $w_{1,i} = \sqrt{|\xi_{1,i}|}(i = 1,2)$.

Now, $|v| \le 2u$, $|f_2(t,u,v)| \le 2|f_1(t,u,v)|$, and $N = N_1N_2N_3$. Thus, one can easily get that (H1) holds by choosing $N_3 \ge \max\{2, \frac{2}{N_1N_2}\}$, where $N_j = \sup\limits_{0<t,s<1} \frac{G_{2,j}(t,s)}{G_{1,j}(t,s)}$, $j = 1,2$.

In addition, by calculation, we get that

$$\lim_{\substack{|v|\le Nu \\ u\to 0^+}} \sup \max_{t\in[0,1]} \frac{f_1(t,u,v)}{u} = \lim_{\substack{|v|\le Nu \\ u\to 0^+}} \sup \max_{t\in[0,1]} \frac{(\pi^4 - 2\pi^2)(1+\sin(\pi t))uv^{\frac{1}{4}}}{u} = 0 < \lambda_1,$$

$$\lim_{\substack{|v|\le Nu \\ u\to +\infty}} \sup \max_{t\in[0,1]} \frac{f_1(t,u,v)}{u} = \lim_{\substack{|v|\le Nu \\ u\to +\infty}} \sup \max_{t\in[0,1]} \frac{(\pi^4 - 2\pi^2)(1+\sin(\pi t))u^{\frac{1}{4}}v^{\frac{1}{4}}}{u} = 0 < \lambda_1.$$

Choose

$$r = \min\{1, [\delta_{1,1}\delta_{1,2} \int_0^1 \sqrt{\sigma(t)} \sin(\pi t) dt \min_{t\in[\frac{1}{4},\frac{3}{4}]} (G_{1,1}(t,t)G_{1,2}(t,t))^2]^2\} > 0$$

and

$$\Phi_r(t) = \sqrt{r\sigma(t)} \sin(\pi t).$$

Then, it is not difficult to obtain that the condition (2) in Theorem 3 holds. Hence, our conclusion follows from Theorem 3. □

Example 2. *Consider the following BVP of fourth-order ordinary differential systems*

$$\begin{cases} u^{(4)}(t) + u''(t) - \pi^2 u(t) = f_1(t,u,v), \ 0 < t < 1; \\ v^{(4)}(t) + \frac{1}{2}v''(t) - \frac{\pi^2}{2}v(t) = f_2(t,u,v), \ 0 < t < 1; \\ u(0) = u(1) = u''(0) = u''(1) = 0; \\ v(0) = v(1) = v''(0) = v''(1) = 0. \end{cases} \tag{28}$$

where

$$f_1(t,u,v) = \begin{cases} (\pi^4 - 2\pi^2)(2+t)u^{\frac{1}{2}}v^{\frac{1}{3}} & \text{if } 0 < u < 1, \ 0 < v < 1; \\ (\pi^4 - 2\pi^2)(2+t)v^{\frac{1}{3}} & \text{if } u = 1, \quad 0 < v < 1; \\ (\pi^4 - 2\pi^2)(2+t)u^2 v^{\frac{1}{3}} & \text{if } u > 1, \quad 0 < v < 1, \end{cases}$$

$$f_2(t,u,v) = \begin{cases} (\pi^4 - 2\pi^2)(1+\cos(\pi t))u^{\frac{1}{2}}v^{\frac{1}{3}} & \text{if } 0 < u < 1, \ 0 < v < 1; \\ (\pi^4 - 2\pi^2)(1+\cos(\pi t))v^{\frac{1}{3}} & \text{if } u = 1, \quad 0 < v < 1; \\ (\pi^4 - 2\pi^2)(1+\cos(\pi t))u^2 v^{\frac{1}{3}} & \text{if } u > 1, \quad 0 < v < 1, \end{cases}$$

Then, BVP (28) has at least two nontrivial solutions.

Proof of Example 2 . BVP (28) can be regarded as a BVP of the form (1). Using a similar process of the proof of Example 1, one can easily obtain that

$$\liminf_{\substack{|v|\le Nu \\ u\to 0^+}} \min_{t\in[0,1]} \frac{f_1(t,u,v)}{u} = \liminf_{\substack{|v|\le Nu \\ u\to 0^+}} \min_{t\in[0,1]} \frac{(\pi^4 - 2\pi^2)(2+t)u^{\frac{1}{2}}v^{\frac{1}{3}}}{u} = +\infty > \pi^4 - 2\pi^2 = \lambda_1,$$

$$\liminf_{\substack{|v|\le Nu \\ u\to +\infty}} \min_{t\in[0,1]} \frac{f_1(t,u,v)}{u} = \liminf_{\substack{|v|\le Nu \\ u\to +\infty}} \min_{t\in[0,1]} \frac{(\pi^4 - 2\pi^2)(2+t)u^2 v^{\frac{1}{3}}}{u} + \infty > \pi^4 - 2\pi^2 = \lambda_1.$$

In addition, it is obvious that (H1) holds by choosing $N_3 = 2$. In the following, set

$$R = \max\{1, \frac{2}{5\pi^2 C_{1,1}C_{1,2} \max_{t\in[0,1]}[G_{1,1}(t,t)G_{1,2}(t,t)]}\} > 0$$

and

$$\Psi_R(t) = \pi^4 R^2 (2+t).$$

Then, it is trivial to verify that assumption (2) of Theorem 3 is true.

As a result, by Theorem 4, system (28) has at least two nontrivial solutions. □

5. Conclusions

In this paper, we have obtained some appropriate results corresponding to multiple solutions for a class of nonlinear fourth-order boundary value problems with parameters. The multiple solutions for the considered systems are obtained under some suitable assumptions via fixed point index theory. The whole theoretical results has been demonstrated by providing two interesting examples. Hence, we claim that fixed point index theory can be used as a strong technique to study nonlinear fourth-order boundary value problems with parameters.

Author Contributions: Conceptualization and Visualization, Y.L.; Formal analysis and Investigation, D.Z.; Writing original draft and Investigation, L.L. All authors have read and agreed to the published version of the manuscript.

Funding: This research was funded by NNSF of P.R.China (62073202 and 11671237), and a project of Shandong Province Higher Educational Science and Technology Program of China under the grant J18KA233.

Acknowledgments: The authors are thankful to the editor and anonymous referees for their valuable comments and suggestions.

Conflicts of Interest: The authors declare no conflict of interest.

References

1. Ji, J.; Yang, B. Eigenvalue comparisons for boundary value problems for second order difference equations. *J. Math. Anal. Appl.* **2006**, *320*, 964–972. [CrossRef]
2. Liang, R.; Shen, J. Periodic boundary value problem for second-order impulsive functional differential equations. *Appl. Math. Comput.* **2007**, *193*, 560–571. [CrossRef]
3. Niu, Y.; Yan, B. The existence of positive solutions for the singular two-point boundary value problem. *Topol. Methods Nonlinear Anal.* **2017**, *49*, 665–682. [CrossRef]
4. Pino, M.A.D.; Manasevich, R.F. Existence for a fourth-order boundary value problem under a two-parameter nonresonance condition. *Proc. Am. Math. Soc.* **1991**, *112*, 81–86. [CrossRef]
5. Liu, Y.; O'Regan, D. Multiplicity results for a class of fourth order semipositone m-point boundary value problems.*Appl. Anal.* **2012**, *91*, 911–921. [CrossRef]
6. Zhang, X.; Liu, L. Eigenvalue of fourth-order m-point boundary value problem with derivatives. *Comput. Math. Appl.* **2008**, *56*, 172–185. [CrossRef]
7. Cabada, A.; Wang, G. Positive solutions of nonlinear fractional differential equations with integral boundary value conditions. *J. Math. Anal. Appl.* **2012**, *389*, 403–411. [CrossRef]
8. Liu, B.; Liu, Y. Positive solutions of a two-point boundary value problem for singular fractional differential equations in Banach space. *J. Funct. Space* **2013**, *2013*, 585639. [CrossRef]
9. Liu, Y. Positive solutions using bifurcation techniques for boundary value problems of fractional differential equations. *Abstr. Appl. Anal.* **2013**, *2013*, 162418. [CrossRef]
10. Liu, Y. Bifurcation techniques for a class of boundary value problems of fractional impulsive differential equations. *J. Nonlinear Sci. Appl.* **2015**, *8*, 340–353. [CrossRef]
11. Ma, T.; Yan, B. The multiplicity solutions for nonlinear fractional differential equations of Riemann-Liouville type. *Fract. Calc. Appl. Anal.* **2018**, *21*, 801–818. [CrossRef]
12. Aftabizadeh, A.R. Existence and uniqueness theorems for fourth-order boundary value problems. *J. Math. Anal. Appl.* **1986**, *116*, 415–426. [CrossRef]
13. Hussain, N.; Taoudi, M.A. Krasnosel'skii-type fixed point theorems with applications to Volterra integral equations. *Fixed Point Theory A* **2013**, *2013*, 196. [CrossRef]
14. Lu, H. Multiple Positive Solutions for Singular Semipositone Periodic Boundary Value Problems with Derivative Dependence. *J. Appl. Math.* **2012**, *2012*, 295209. [CrossRef]
15. Lu, H.; Wang, Y.; Liu, Y. Nodal Solutions for Some Second-Order Semipositone Integral Boundary Value Problems. *Abstr. Appl. Anal.* **2014**, *2014*, 951824. [CrossRef]
16. Mao, J.; Zhao, D. Multiple positive solutions for nonlinear fractional differential equations with integral boundary value conditions and a parameter. *J. Funct. Space.* **2019**, *2019*, 2787569. [CrossRef]
17. Ragusa, M.A.; Razani, A. Weak solutions for a system of quasilinear elliptic equations. *Contrib. Math.* **2020**, *1*, 1116. [CrossRef]

18. Ragusa, M.A.; Tachikawa, A. Regularity for minimizers for functionals of double phase with variable exponents. *Adv. Nonlinear Anal.* **2020**, *9*, 710–728. [CrossRef]
19. Simon, L. On qualitative behavior of multiple solutions of quasilinear parabolic functional equations. *Electron. J. Qual. Theory Differ. Equ.* **2020**, *32*, 1–11. [CrossRef]
20. Suo, J.; Wang, W. Eigenvalues of a class of regular fourth-order Sturm-Liouville problems. *Appl. Math. Comput.* **2012**, *218*, 9716–9729. [CrossRef]
21. Vong, S. Positive solutions of singular fractional differential equations with integral boundary conditions. *Math. Comput. Model.* **2013**, *57*, 1053–1059. [CrossRef]
22. Wang, Y.; Liu, Y.; Cui, Y. Infinitely many solutions for impulsive fractional boundary value problem with p-Laplacian. *Bound. Value Probl.* **2018**, *2018*, 94. [CrossRef]
23. Yan, B. Positive solutions for the singular nonlocal boundary value problems involving nonlinear integral conditions. *Bound. Value Probl.* **2014**, *2014*, 38. [CrossRef]
24. Yan, B.; O'Regan, D.; Agarwal, R.P. Positive solutions for singular nonlocal boundary value problems. *Dynam. Syst.* **2014**, *29*, 301–321. [CrossRef]
25. Zhou, B.; Zhang, L.; Addai, E.; Zhang, N. Multiple positive solutions for nonlinear high-order Riemann-Liouville fractional differential equations boundary value problems with p-Laplacian operator. *Bound. Value Probl.* **2020**, *2020*, 26. [CrossRef]
26. Zhao, D.; Liu, Y. Positive solutions for a class of fractional differential coupled system with integral boundary value conditions. *J. Nonlinear Sci. Appl.* **2016**, *9*, 2922–2942. [CrossRef]
27. Zhao, D.; Liu, Y. Twin solutions to semipositone boundary value problems for fractional differential equations with coupled integral boundary conditions. *J. Nonlinear Sci. Appl.* **2017**, *10*, 3544–3565. [CrossRef]
28. Zhao, D.; Liu, Y. Eigenvalues of a class of singular boundary value problems of impulsive differential equations in Banach spaces. *J. Funct. Space* **2014**, *2014*, 720494. [CrossRef]
29. Zhao, D.; Liu, Y.; Li, X. Controllability for a class of semilinear fractional evolution systems via resolvent operators. *Commun. Pur. Appl. Anal.* **2019**, *18*, 455–478. [CrossRef]
30. Agarwal, R. On fourth-order boundary value problems arising in beam analysis. *Differ. Integral Equ.* **1989**, *2*, 91–110.
31. Coster, C.D.; Fabry, C.; Munyamarere, F. Nonresonance conditions for fourth-order nonlinear boundary value problems, Internat. *J. Math. Math. Sci.* **1994**, *17*, 725–740. [CrossRef]
32. Li, Y. Positive solutions of fourth-order boundary value problems with two parameters. *J. Math. Anal. Appl.* **2003**, *281*, 477–484. [CrossRef]
33. Wang, Q.; Lu, Y. Positive solutions for a nonlinear system of fourth-order ordinary differential equations. *Electron. J. Differ. Equ.* **2020**, *45*, 1–15.
34. Guo, D.; Lakshmikantham, V. *Nonlinear Problems in Abstract Cones;* Academic Press: New York, NY, USA, 1988.

Publisher's Note: MDPI stays neutral with regard to jurisdictional claims in published maps and institutional affiliations.

Article

Existence of Three Solutions for a Nonlinear Discrete Boundary Value Problem with ϕ_c-Laplacian

Yanshan Chen [1,2] and Zhan Zhou [1,2,*]

[1] School of Mathematics and Information Science, Guangzhou University, Guangzhou 510006, China; 2111815047@e.gzhu.edu.cn

[2] Center for Applied Mathematics, Guangzhou University, Guangzhou 510006, China

* Correspondence: zzhou@gzhu.edu.cn or zzhou0321@hotmail.com

Received: 14 October 2020; Accepted: 3 November 2020; Published: 6 November 2020

Abstract: In this paper, based on critical point theory, we mainly focus on the multiplicity of nontrivial solutions for a nonlinear discrete Dirichlet boundary value problem involving the mean curvature operator. Without imposing the symmetry or oscillating behavior at infinity on the nonlinear term f, we respectively obtain the sufficient conditions for the existence of at least three non-trivial solutions and the existence of at least two non-trivial solutions under different assumptions on f. In addition, by using the maximum principle, we also deduce the existence of at least three positive solutions from our conclusion. As far as we know, our results are supplements to some well-known ones.

Keywords: ϕ_c-Laplacian; boundary value problem; critical point theory; three solutions

---

1. Introduction

Let $\mathbb{Z}$ and $\mathbb{R}$ denote all integers and real numbers, respectively. Let N be a fixed positive integer. Define $\mathbb{Z}(a,b) = \{a, a+1, \cdots, b\}$ with $a \leq b$ for any $a, b \in \mathbb{Z}$.

Difference equations are widely used in various research fields, such as computer science, discrete optimization, economics and biological neural networks [1–4]. On the existence and multiplicity of solutions for the boundary value problems of difference equations, many authors have come to important conclusions by exploiting various methods, including the method of upper and lower solutions, Brouwer degree and invariant sets of descending flow [5–7]. Critical point theory was used largely to explore differential equations much earlier on in history. In 2003, Guo and Yu in [8] used critical point theory for the first time to obtain sufficient conditions for the existence of periodic solutions and subharmonic solutions of difference equations. This crucial breakthrough inspired many scholars to use critical point theory to study the dynamics of difference equations and many meaningful and interesting results have been obtained, especially in periodic solutions [9–11], homoclinic solutions [12–16] and boundary value problems [17–23].

In [24], Agarwal, Perera and O'Regan employed the critical point theory to establish the existence of at least two positive solutions of the following second order discrete boundary value problem

$$\begin{cases} -\Delta^2 x(k-1) = f(k, x(k)), \ k \in \mathbb{Z}(1, N), \\ x(0) = x(N+1) = 0, \end{cases}$$

where $\Delta x(k) = x(k+1) - x(k)$, $\Delta^2 x(k) = \Delta(\Delta x(k))$ and $f : \mathbb{Z}(1,N) \times \mathbb{R} \to \mathbb{R}$ is a continuous function. In [18], by using the three critical point theorems proposed by Bonanno [25], Jiang and Zhou obtained sufficient conditions for the existence of at least three solutions of the following Dirichlet boundary value problem with ϕ_p-Laplacian

$$\begin{cases} -\Delta(\phi_p(\Delta u(k-1))) = \lambda f(k,u(k)),\ k \in \mathbb{Z}(1,N), \\ u(0) = u(N+1) = 0, \end{cases}$$

where ϕ_p is the p-Laplacian defined by $\phi_p(u) = |u|^{p-2}u$ $(p > 1)$ with $u \in \mathbb{R}$ and $\lambda > 0$ is a positive parameter. Different from the conclusion of [18], Bonanno in [26] obtained the existence of three positive solutions without the asymptotic condition of the nonlinear function f. In particular, Bonanno obtained the sufficient conditions for the existence of at least four nontrivial solutions when f satisfies the growth condition at zero and infinity [26], which improved the result in [18].

In [27], by using critical point theory, Nastasi and Vetro obtained the existence of at least two positive solutions to the following Dirichlet boundary value problem with (p,q)-Laplacian

$$\begin{cases} -\Delta(\phi_p(\Delta u(z-1))) - \Delta(\phi_q(\Delta u(z-1))) + \alpha(z)\phi_p(u(z)) + \beta(z)\phi_q(u(z)) = \lambda g(z,u(z)),\ z \in \mathbb{Z}(1,N), \\ u(0) = u(N+1) = 0, \end{cases}$$

where $1 < q < p < +\infty$, α, $\beta : \mathbb{Z}(1,N) \to \mathbb{R}$ and $g : [1,N+1] \times \mathbb{R} \to \mathbb{R}$ are a continuous function with $g(N+1,t) = 0$ for all $t \in \mathbb{R}$.

It is well known that the differential equation with ϕ_c-Laplacian (ϕ_c is the mean curvature operator defined by $\phi_c(s) = s/\sqrt{1+s^2}$ for $s \in \mathbb{R}$) was studied by many scholars in the past decades [28–31]. It is usually regarded as a variant of the Liouville–Bratu–Gelfand problem, which is used to study the dynamic model of combustible gases. As mentioned above, we find that the research on the difference equation largely focuses on the case with ϕ_p-Laplacian. However, there are only a few results on the boundary value problems involving ϕ_c-Laplacian [32]. Recently, Zhou and Ling in [33] considered the existence of multiple solutions of the following discrete Dirichlet boundary value problem with ϕ_c-Laplacian.

Problem 1

$$\begin{cases} -\Delta\left(\frac{\Delta u(k-1)}{\sqrt{1+(\Delta u(k-1))^2}}\right) = \lambda f(k,u(k)),\ k \in \mathbb{Z}(1,N), \\ u(0) = u(N+1) = 0. \end{cases}$$

The authors in [33] found that the properties of the nonlinear term $F(k,u) = \int_0^u f(k,t)dt$ plays an important role on the existence of multiple solutions. When $F(k,u)$ has oscillating behavior at infinity, there are infinite solutions to the boundary value (Problem 1) [33]. Naturally, we would like to ask: What will happen if $F(k,u)$ does not oscillate at infinity?

To address this problem, in this paper, we will study the existence of solutions for the boundary value (Problem 1) without oscillating nonlinear terms. In fact, based on the theorems of G. Bonanno (Theorem 4.1 in [22] and Theorem 2.1 in [23]), we will give the conditions of the existence of at least three nontrivial solutions for (Problem 1), when $F(k,u)$ does not have oscillation property at infinity. In addition, when $f(k,0) > 0$, we obtain the sufficient conditions for the existence of at least three positive solutions of (Problem 1). Moveover, we give two examples to illustrate our main results.

For convenience, we end this section by recalling some classical definitions and two well-known lemmas, which are the main tools of this paper.

Let $(X, \|\cdot\|)$ be a real Banach space. We say that $I : X \to \mathbb{R}$ is coercive on X if $\lim_{\|u\|\to+\infty} I(u) = +\infty$. If I is a continuously Gâteaux differentiable functional, we say that I satisfies the Palais–Smale condition ((PS)-condition in short), if any sequence $u_n \subset X$ such that $\{I(u_n)\}$ is bounded and $\{I'(u_n)\}$ is convergent to 0 in X^*, has a convergent subsequence in X.

(H) Let $(X, \|\cdot\|)$ *be a real finite dimensional Banach space and let* $\Phi, \Psi : X \to \mathbb{R}$ *be two continuously Gâteaux differentiable functionals with* Φ *coercive and such that*

$$\inf_X \Phi = \Phi(0) = \Psi(0) = 0.$$

Lemma 1 ([22]). *Assume that (H) holds and there exist* $r > 0$ *and* $\overline{x} \in X$, *with* $0 < r < \Phi(\overline{x})$, *such that:*

(a_1) $\dfrac{\sup_{\Phi(x)\leq r} \Psi(x)}{r} < \dfrac{\Psi(\overline{x})}{\Phi(\overline{x})}$;

(a_2) *for each* $\lambda \in \Lambda_r := \left(\dfrac{\Phi(\overline{x})}{\Psi(\overline{x})}, \dfrac{r}{\sup_{\Phi(x)\leq r} \Psi(x)} \right)$, *the functional* $\Phi - \lambda\Psi$ *is coercive.*

Then, for each $\lambda \in \Lambda_r$, *the functional* $I_\lambda = \Phi - \lambda\Psi$ *has at least three distinct critical points in X.*

Lemma 2 ([23]). *Let X be a real Banach space and let* $\Phi, \Psi : X \to \mathbb{R}$ *be two continuously Gâteaux differentiable functionals such that* $\inf_X \Phi = \Phi(0) = \Psi(0) = 0$. *Assume that there are* $r > 0$ *and* $\tilde{u} \in X$, *with* $0 < \Phi(\tilde{u}) < r$, *such that*

$$\frac{\sup_{\Phi(u)\leq r} \Psi(u)}{r} < \frac{\Psi(\tilde{u})}{\Phi(\tilde{u})} \tag{1}$$

for each $\lambda \in \left(\dfrac{\Phi(\tilde{u})}{\Psi(\tilde{u})}, \dfrac{r}{\sup_{\Phi(u)\leq r} \Psi(u)} \right)$, *and the functional* $I_\lambda = \Phi - \lambda\Psi$ *satisfies (PS)-condition and it is unbounded from below. Then, for each* $\lambda \in \left(\dfrac{\Phi(\tilde{u})}{\Psi(\tilde{u})}, \dfrac{r}{\sup_{\Phi(u)\leq r} \Psi(u)} \right)$, *the functional* I_λ *admits at least two non-zero critical points* $u_{\lambda,1}, u_{\lambda,2}$ *such that* $I_\lambda(u_{\lambda,1}) < 0 < I_\lambda(u_{\lambda,2})$.

2. Preliminaries

In this section, we recall some definitions, notations and properties. Consider the N-dimensional Banach space

$$S = \{u : [0, N+1] \to \mathbb{R} : u(0) = u(N+1) = 0\}$$

endowed with the norm

$$\|u\| = \left(\sum_{k=1}^{N+1} |\Delta u(k-1)|^2 \right)^{\frac{1}{2}}.$$

We define the functional $\Phi, \Psi : S \to \mathbb{R}$ in the following way

$$\Phi(u) = \sum_{k=1}^{N+1} \left(\sqrt{1 + (\Delta u(k-1))^2} - 1 \right), \quad \Psi(u) = \sum_{k=1}^{N} F(k, u(k)) \tag{2}$$

for each $u \in S$, where

$$F(k, \xi) := \int_0^\xi f(k, t)dt, \quad \xi \in \mathbb{R}.$$

It is easy to check that $\Phi, \Psi \in C^1(S, R)$ and we have

$$\Phi'(u)(v) = -\sum_{k=1}^{N} \Delta(\phi_c(\Delta u(k-1)))v(k)$$

and

$$\Psi'(u)(v) = -\sum_{k=1}^{N} f(k, u(k))v(k).$$

By a standard argument, it can be shown that the critical points of the functional

$$I_\lambda = \Phi - \lambda\Psi$$

are the solutions of (Problem 1).

Let

$$\|u\|_\infty = \max\left\{|u(k)| : k \in \mathbb{Z}(1,N)\right\}.$$

We see that $\|\cdot\|_\infty$ is another norm in S. From Lemma 2.2 of [18], we have

Lemma 3 ([18]). *For any $u \in S$, the following relation holds*

$$\|u\|_\infty \leq \frac{\sqrt{N+1}}{2}\|u\|.$$

From (2.1) and (2.3) in [22], we have

Lemma 4 ([22]). *For any $u \in S$, one has*

$$\frac{1}{\sqrt{N\lambda_N}}\|u\| \leq \|u\|_\infty \leq \frac{1}{\sqrt{\lambda_1}}\|u\|,$$

where $\lambda_1 = 4\sin^2\frac{\pi}{2(N+1)}$ and $\lambda_N = 4\sin^2\frac{N\pi}{2(N+1)}$.

Finally, in order to obtain the positive solutions of (Problem 1), we need the following strong maximum principle, which can be found in Theorem 2.1 of [33].

Lemma 5 ([33]). *Assume $u \in S$ such that either*

$$u(k) > 0 \quad or \quad \Delta(\phi_c(\Delta u(k-1))) \leq 0,$$

for all $k \in \mathbb{Z}(1,N)$. Then, either $u > 0$ in $\mathbb{Z}(1,N)$ or $u \equiv 0$.

3. Main Results

For convenience, set

$$F_t := \sum_{k=1}^{N} F(k,t), \text{ for all } t > 0.$$

Our first result is the following theorem.

Theorem 1. *Assume that there exist two positive constants c and d with*

$$2\sqrt{1+d^2} > 1 + \sqrt{1+\frac{4c^2}{N+1}} \tag{3}$$

such that

(i) $f(k,\xi) > 0$ *for each* $k \in \mathbb{Z}(1,N)$ *and* $\xi \in [-c,c]$;

(ii) $\frac{F_d}{2(\sqrt{1+d^2}-1)} > \frac{F_c}{-1+\sqrt{1+\frac{4c^2}{N+1}}}$;

(iii) $\limsup\limits_{|\xi|\to+\infty}\frac{F(k,\xi)}{|\xi|} < \frac{2F_c}{N(\sqrt{4c^2+N+1}-\sqrt{N+1})}$.

Then, for every $\lambda \in \Lambda := \left(\frac{2(\sqrt{1+d^2}-1)}{F_d}, \frac{-1+\sqrt{1+\frac{4c^2}{N+1}}}{F_c}\right)$, *(Problem 1) has at least three nontrivial solutions.*

Proof. We take $X = S$, Φ and Ψ as in (2). Clearly, Φ and Ψ are two continuously Gâteaux differentiable functionals. Now, we prove the coercivity of Φ. In fact, one has

$$\begin{aligned}\Phi(u) &= \sum_{k=1}^{N+1}\left(\sqrt{1+(\Delta u(k-1))^2}-1\right)\\ &\geq \left(\sum_{k=1}^{N+1}\left[1+(\Delta u(k-1))^2\right]\right)^{\frac{1}{2}} - \sum_{k=1}^{N+1} 1\\ &\geq \|u\| - N - 1.\end{aligned}$$

This means $\lim_{\|u\|\to+\infty} \Phi(u) = +\infty$ and verifies the coercivity of Φ. Moreover, by the definition of Φ and Ψ, we can obtain

$$\inf_X \Phi = \Phi(0) = \Psi(0) = 0.$$

To summarize, condition (H) in Lemma 1 holds.

According to Lemma 1, it is clear that Theorem 1 holds if we can verify (a_1) and (a_2) of Lemma 1. Put

$$r = -1 + \sqrt{1+\frac{4c^2}{N+1}}.$$

If $\Phi(u) \leq r$, let

$$v(k) = \sqrt{1+(\Delta u(k))^2} - 1 \text{ for } k \in \mathbb{Z}(0, N).$$

Then, we have $\sum_{k=0}^{N} v(k) \leq r$ and

$$\sum_{k=0}^{N}(\Delta u(k))^2 = \sum_{k=0}^{N}\left(v(k)^2 + 2v(k)\right) \leq \left(\sum_{k=0}^{N} v(k)\right)^2 + 2\sum_{k=0}^{N} v(k) \leq r^2 + 2r = \frac{4c^2}{N+1}.$$

Thus, $\|u\|_\infty \leq \frac{\sqrt{N+1}}{2}\|u\| \leq c$ by Lemma 3.

By (i), we see that $F(k,\xi)$ is increasing in $\xi \in [-c, c]$. Thus

$$\frac{\sup_{\Phi(u)\leq r} \Psi(u)}{r} \leq \frac{\sup_{\|u\|_\infty\leq c} \sum_{k=1}^{N} F(k, u(k))}{-1+\sqrt{1+\frac{4c^2}{N+1}}} \leq \frac{\sum_{k=1}^{N} \max_{|\xi|\leq c} F(k,\xi)}{-1+\sqrt{1+\frac{4c^2}{N+1}}} = \frac{F_c}{-1+\sqrt{1+\frac{4c^2}{N+1}}}. \tag{4}$$

Then, it is easy to get

$$\frac{\sup_{\Phi(u)\leq r} \Psi(u)}{r} < \frac{1}{\lambda}. \tag{5}$$

Now, let $\overline{u} \in S$ be defined by

$$\overline{u}(k) = \begin{cases} d, & \text{if } k \in \mathbb{Z}(1, N), \\ 0, & \text{if } k = 0 \text{ or } k = N+1. \end{cases}$$

Then, we see from (3) that

$$\Phi(\overline{u}) = 2(\sqrt{1+d^2} - 1) > \sqrt{1+\frac{4c^2}{N+1}} - 1 = r.$$

Moreover, it holds that

$$\frac{\Psi(\overline{u})}{\Phi(\overline{u})} = \frac{\sum_{k=1}^{N} F(k, \overline{u}(k))}{2(\sqrt{1+d^2}-1)} = \frac{F_d}{2(\sqrt{1+d^2}-1)}. \tag{6}$$

Therefore, we have

$$\frac{\Psi(\overline{u})}{\Phi(\overline{u})} > \frac{1}{\lambda}. \tag{7}$$

Hence, condition (a_1) of Lemma 1 follows by combining (5) with (7).

Next, we prove the coercivity of the functional I_λ. From (iii), there is an $\varepsilon > 0$ such that

$$\limsup_{|\xi|\to+\infty} \frac{F(k,\xi)}{|\xi|} < \varepsilon < \frac{2F_c}{N(\sqrt{4c^2+N+1}-\sqrt{N+1})}.$$

Then, there is a positive constant h such that

$$F(k,\xi) \le \varepsilon|\xi| + h$$

for each $\xi \in \mathbb{R}$ and $k \in \mathbb{Z}(1,N)$. By Lemma 3 and $\lambda < \frac{-1+\sqrt{1+\frac{4c^2}{N+1}}}{F_c}$, one has

$$\begin{aligned}
\lambda \sum_{k=1}^{N} F(k,u(k)) &\le \lambda \sum_{k=1}^{N} [\varepsilon|u(k)| + h] \\
&\le \lambda \frac{\varepsilon N\sqrt{N+1}}{2}\|u\| + \lambda N h \\
&\le \frac{\varepsilon N\left(\sqrt{N+1+4c^2}-\sqrt{N+1}\right)}{2F_c}\|u\| + \frac{\left(-1+\sqrt{1+\frac{4c^2}{N+1}}\right)Nh}{F_c},
\end{aligned}$$

for each $u \in S$. Therefore, it is true that

$$\begin{aligned}
I_\lambda(u) &= \Phi(u) - \lambda\Psi(u) \\
&= \sum_{k=1}^{N+1}\left[\sqrt{1+(\Delta u(k-1))^2}-1\right] - \lambda\sum_{k=1}^{N} F(k,u(k)) \\
&\ge \left[\sum_{k=1}^{N+1}\left[1+(\Delta u(k-1))^2\right]\right]^{\frac{1}{2}} - N - 1 - \lambda\sum_{k=1}^{N} F(k,u(k)) \\
&\ge \left[1 - \frac{\varepsilon N\left(\sqrt{N+1+4c^2}-\sqrt{N+1}\right)}{2F_c}\right]\|u\| - N - 1 - \frac{\left(-1+\sqrt{1+\frac{4c^2}{N+1}}\right)Nh}{F_c}.
\end{aligned}$$

Thus, we get that I_λ is coercive, and condition (a_2) of Lemma 1 is verified. In summary, all assumptions of Lemma 1 are proved, and so the functional I_λ has at least three distinct critical points in X for each $\lambda \in \Lambda$. Since $u \equiv 0$ is not a solution to the (Problem 1), the proof is completed. □

Remark 1. *It is obvious that the mean curvature operator ϕ_c is odd symmetric ($\phi_c(-s) = -\phi_c(s)$, $s \in \mathbb{R}$). When the nonlinear term f is also odd symmetric ($f(\cdot,-s) = -f(\cdot,s)$), the variational functional I_λ is even symmetric ($I_\lambda(-u) = I_\lambda(u)$, $u \in S$). In this case, it is easy to obtain multiple solutions to (Problem 1) by using the critical point theory with symmetries. However, in this paper, we obtain multiple solutions to (Problem 1) without the symmetry on f.*

Now, let

$$F^+(k,\xi) = \int_0^\xi f(k,t^+)dt, \quad (k,\xi) \in \mathbb{Z}(1,N) \times \mathbb{R},$$

where $t^+ = \max\{0,t\}$ and define $I_\lambda^+ = \Phi - \lambda\Psi^+$, with Φ defined in (2) and

$$\Psi^+(u) := \sum_{k=1}^{N} F^+(k,u(k)).$$

It is well known that $I_\lambda^+ \in C^1(S,\mathbb{R})$ and the critical points of I_λ^+ are precisely the solutions of the following problem.

Problem 2

$$\begin{cases} -\Delta(\phi_c(\Delta u(k-1))) = \lambda f(k,u^+(k)), \ k \in \mathbb{Z}(1,N), \\ u(0) = u(N+1) = 0. \end{cases}$$

We have the following corollary.

Corollary 1. *Assume that there exist two positive constants c and d with*

$$2\sqrt{1+d^2} > 1 + \sqrt{1 + \frac{4c^2}{N+1}}$$

such that

(i) $f(k,\xi) > 0$ *for each* $k \in \mathbb{Z}(1,N)$ *and* $\xi \in [0,c]$;

(ii) $\frac{F_d}{2(\sqrt{1+d^2}-1)} > \frac{F_c}{-1+\sqrt{1+\frac{4c^2}{N+1}}}$;

(iii) $\limsup_{\xi\to+\infty} \frac{F(k,\xi)}{\xi} < \frac{2F_c}{N(\sqrt{4c^2+N+1}-\sqrt{N+1})}$.

Then, for every $\lambda \in \Lambda$, *(Problem 1) has at least three positive solutions.*

Proof. For each $k \in \mathbb{Z}(1,N)$, consider (Problem 2) with

$$f^+(k,\xi) = \begin{cases} f(k,\xi), & \text{if } \xi > 0, \\ f(k,0), & \text{if } \xi \le 0. \end{cases}$$

One has that condition (i) of Theorem 1 holds. Besides, we have

$$\limsup_{\xi\to-\infty} \frac{F^+(k,\xi)}{|\xi|} = \limsup_{\xi\to-\infty} \frac{\xi f(k,0)}{|\xi|} = -f(k,0) < \frac{2F_c}{N(\sqrt{4c^2+N+1}-\sqrt{N+1})}.$$

So, all conditions of Theorem 1 are true. Moreover, since $u \equiv 0$ is not a solution of problem $(P_{\lambda,f}^+)$, we can get that (Problem 2) has at least three nontrivial solutions. Assume $u = \{u(k)\}$ is one of the nontrivial solution, for any $k \in \mathbb{Z}(1,N)$, one has either $u(k) > 0$ or

$$-\Delta(\phi_c(\Delta u(k-1))) = \lambda f(k,u^+(k)) = \lambda f(k,0) > 0.$$

Then, we have $u > 0$ for all $k \in \mathbb{Z}(1,N)$ by Lemma 5, i.e., u is a positive solution. In addition, if u is a positive solution of (Problem 2), then u is a positive solution of (Problem 1) obviously and Corollary 1 is proved. □

Next, we will use Lemma 2 to obtain another conclusion of this paper.

Theorem 2. *Assume that there exist two positive constants c and d with*

$$2\sqrt{1+d^2} < 1 + \sqrt{1 + \frac{4c^2}{N+1}} \tag{8}$$

such that

(T_1) $f(k,\xi) > 0$ *for each* $k \in \mathbb{Z}(1,N)$ *and* $\xi \in [-c,c]$;

(T_2) $\max\left\{ \frac{2(\sqrt{1+d^2}-1)}{F_d}, \frac{2\sqrt{N(N+1)}\sin\frac{N\pi}{2(N+1)}}{\beta} \right\} < \frac{-1+\sqrt{1+\frac{4c^2}{N+1}}}{F_c}$;

(T_3) *there exist a positive constant* β *such that*

$$\liminf_{|t|\to+\infty} \frac{F(k,t)}{|t|} > \beta$$

for each $k \in \mathbb{Z}(1,N)$.

Then, for every $\lambda \in \left(\max\left\{ \frac{2(\sqrt{1+d^2}-1)}{F_d}, \frac{2\sqrt{N(N+1)}\sin\frac{N\pi}{2(N+1)}}{\beta} \right\}, \frac{-1+\sqrt{1+\frac{4c^2}{N+1}}}{F_c} \right)$, *(Problem 1) has at least two non-zero critical points* $u_{\lambda,1}, u_{\lambda,2}$ *such that* $I_\lambda(u_{\lambda,1}) < 0 < I_\lambda(u_{\lambda,2})$.

Proof. Clearly, Φ, Ψ are two continuously Gâteaux differentiable functionals and

$$\inf_X \Phi = \Phi(0) = \Psi(0) = 0.$$

Let $\overline{u}$ and r be the same as the ones defined in the proof of Theorem 1. Then we have

$$\frac{\sup\limits_{\Phi(u)\le r} \Psi(u)}{r} \le \frac{F_c}{-1+\sqrt{1+\frac{4c^2}{N+1}}} < \frac{1}{\lambda}$$

and

$$\frac{\Psi(\overline{u})}{\Phi(\overline{u})} = \frac{F_d}{2(\sqrt{1+d^2}-1)} > \frac{1}{\lambda}.$$

So inequality (1) in Lemma 2 holds. Besides, form (T_3) and Lemma 4, there is a constant h such that, for any $u \in S$ and $k \in \mathbb{Z}(1,N)$,

$$\begin{aligned}
\lambda \sum_{k=1}^{N} F(k,u(k)) &\ge \lambda \sum_{k=1}^{N} (\beta|u(k)| - h) \\
&\ge \lambda\beta\|u\|_\infty - \lambda N h \\
&\ge \frac{\lambda\beta}{\sqrt{N\lambda_N}}\|u\| - \lambda N h.
\end{aligned}$$

Therefore, we have

$$\begin{aligned}
I_\lambda(u) &= \Phi(u) - \Psi(u) \\
&= \sum_{k=1}^{N+1}\left[\sqrt{1+(\Delta u(k-1))^2}-1\right] - \lambda\sum_{k=1}^{N}F(k,u(k)) \\
&\leq \sum_{k=1}^{N+1}\left[1+\sqrt{(\Delta u(k-1))^2}-1\right] - \frac{\lambda\beta}{\sqrt{N\lambda_N}}\|u\| + \lambda Nh \\
&\leq \left(\sum_{k=1}^{N+1}1^2\right)^{\frac{1}{2}}\left(\sum_{k=1}^{N+1}(\Delta u(k-1))^2\right)^{\frac{1}{2}} - \frac{\lambda\beta}{\sqrt{N\lambda_N}}\|u\| + \lambda Nh \\
&\leq \left(\sqrt{N+1} - \frac{\lambda\beta}{\sqrt{N\lambda_N}}\right)\|u\| + \lambda Nh \\
&= \left(\sqrt{N+1} - \frac{\lambda\beta}{2\sqrt{N}\sin\frac{N\pi}{2(N+1)}}\right)\|u\| + \lambda Nh.
\end{aligned}$$

Thus, we see that $\lim_{\|u\|\to+\infty} I_\lambda(u) = -\infty$, which means that the functional I_λ is unbounded from below. Moreover, we can get that $-I_\lambda$ is coercive. Therefore, I_λ satisfies the (PS)-condition and the proof is completed. □

4. Examples

Example 1. *Fix $N \in \mathbb{Z}(1,20)$ and consider the boundary value (Problem 1) with*

$$f(k,u) = f(u) = \begin{cases} 3u^2 + \frac{1}{4}\cos u, & \text{if } u \leq \frac{3\pi}{2}, \\ \frac{27\pi^2}{4}\cos(u+\frac{\pi}{2}), & \text{if } u > \frac{3\pi}{2}, \end{cases}$$

for $k \in \mathbb{Z}(1,N)$. Then, we have

$$F(k,\xi) = F(\xi) = \begin{cases} \xi^3 + \frac{1}{4}\sin\xi, & \text{if } \xi \leq \frac{3\pi}{2}, \\ \frac{27\pi^3}{8} - \frac{1}{4} + \frac{27\pi^2}{4}\sin(\xi+\frac{\pi}{2}), & \text{if } \xi > \frac{3\pi}{2}. \end{cases}$$

Let $c = \frac{1}{2}$ and $d = 4$. Then one has

$$2\sqrt{1+d^2} = 2\sqrt{17} > 1 + \sqrt{1+\frac{1}{N+1}} = 1 + \sqrt{1+\frac{4c^2}{N+1}}.$$

In addition, as shown in Figure 1, $f(k,\xi) > 0$ for each $\xi \in [0,\frac{1}{2}]$. Thus condition (i) of Corollary 1 follows. Moreover, we have

$$\frac{F_c}{-1+\sqrt{1+\frac{4c^2}{N+1}}} = \frac{N(1+2\sin\frac{1}{2})}{8\left(-1+\sqrt{1+\frac{1}{N+1}}\right)} \leq \frac{N(1+2\sin\frac{1}{2})}{8\left(-1+\sqrt{\frac{21}{20}}\right)} \approx 9.915N \tag{9}$$

and

$$\frac{F_d}{2(\sqrt{1+d^2}-1)} = \frac{N(256+\sin 4)}{8(\sqrt{17}-1)} \approx 10.216N. \tag{10}$$

Since

$$\frac{256+\sin 4}{8(\sqrt{17}-1)} - \frac{1+2\sin\frac{1}{2}}{8\left(-1+\sqrt{\frac{21}{20}}\right)} > 0,$$

by combining (9) and (10), condition (ii) of Corollary 1 holds.

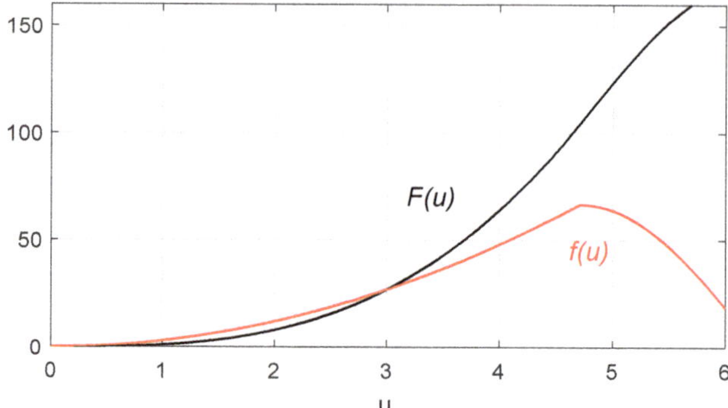

Figure 1. The images of $f(u)$ and $F(u)$ in Example 1.

We can further verify condition (iii) of Corollary 1, since

$$\limsup_{\xi\to+\infty}\frac{F(k,\xi)}{\xi}=\limsup_{\xi\to+\infty}\frac{\frac{27\pi^3}{8}-\frac{1}{4}+\frac{27\pi^2}{4}\sin(\xi+\frac{\pi}{2})}{\xi}=0<\frac{2F_c}{N(\sqrt{4c^2+N+1}-\sqrt{N+1})}.$$

To sum up, all the conditions of Corollary 1 are satisfied. Hence, for

$$\lambda\in\Lambda=\left(\frac{8(\sqrt{17}-1)}{N(256+\sin 4)},\frac{8\left(-1+\sqrt{1+\frac{1}{N+1}}\right)}{N(1+2\sin\frac{1}{2})}\right),$$

the boundary value problem admits at least three positive solutions.

Example 2. *Let $N=3$ and $\beta=10$, consider the boundary value (Problem 1) with*

$$f(k,u)=f(u)=\begin{cases}\frac{1}{6}\left[(3-2u)\cos\frac{u}{2}+e^{u-2}\right], & \text{if } u\le 2\pi,\\ \frac{1}{6}\left(4\pi-3+e^{2\pi-2}\right), & \text{if } u>2\pi,\end{cases}$$

for each $k\in\mathbb{Z}(1,N)$. Then, we have

$$F(k,\xi)=F(\xi)=\begin{cases}\frac{1}{3}\left[(3-2\xi)\sin\frac{\xi}{2}+4\left(1-\cos\frac{\xi}{2}\right)\right]+\frac{e^{-2}}{6}\left(e^{\xi}-1\right), & \text{if } \xi\le 2\pi,\\ \frac{1}{6}\left(4\pi-3+e^{2\pi-2}\right)\xi-\frac{\pi}{3}(4\pi-3+e^{2\pi-2})+\frac{e^{-2}}{6}(e^{2\pi-1})+\frac{8}{3}, & \text{if } \xi>2\pi.\end{cases}$$

Moreover, the images of $f(u)$ and $F(u)$ are shown in Figure 2.
Letting $c=\sqrt{N+1}=2$ and $d=\frac{1}{2}$, we have

$$2\sqrt{1+d^2}=\sqrt{5}<1+\sqrt{5}=1+\sqrt{1+\frac{4c^2}{N+1}}.$$

Now, we prove $f(u)>0$ for each $u\in[-2,2]$. Let

$$f_1(u)=(3-2u)\cos\frac{u}{2}\quad\text{and}\quad f_2(u)=e^{u-2}.$$

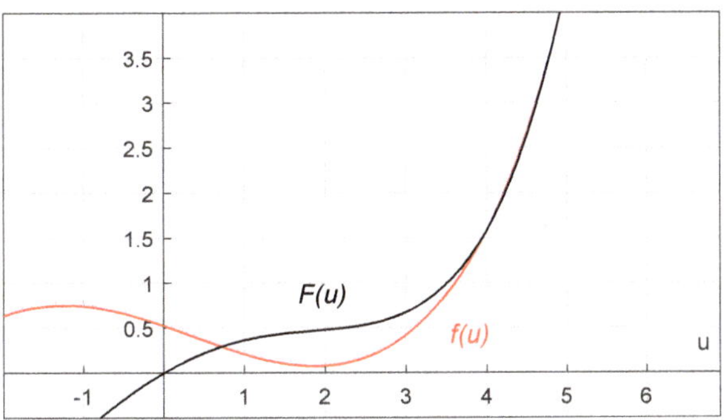

Figure 2. The images of $f(u)$ and $F(u)$ in Example 2.

Then $f(u) = \frac{1}{6}(f_1(u) + f_2(u))$ *for* $u \leq 2\pi$. *Besides, it is obvious that* $f_2(u) > 0$ *on* $[-2, 2]$ *and* $f_1(u) \geq 0$ *on* $\left[-2, \frac{3}{2}\right]$. *When* $u \in \left[\frac{3}{2}, 2\right]$, *we have*

$$f_1'(u) = \frac{1}{2}\left[(2u-3)\sin\frac{u}{2} - 4\cos\frac{u}{2}\right] \leq \frac{1}{2}(\sin 1 - 4\cos 1) < 0.$$

Thus $f_1(u)$ *is decreasing on* $[\frac{3}{2}, 2]$. *Since* $f_2(u)$ *is an increasing function and* $f_1(2) + f_2(\frac{3}{2}) = -\cos 1 + e^{-\frac{1}{2}} > 0$, *we can get* $f(u) > 0$ *on* $[-2, 2]$.

In addition, we have

$$F_d = \sum_{k=1}^{N} F(k,d) = N\int_0^d \frac{1}{6}\left[(3-2u)\cos\frac{u}{2} + e^{u-2}\right] du \approx 0.663$$

and

$$F_c = \sum_{k=1}^{N} F(k,c) = N\int_0^c \frac{1}{6}\left[(3-2u)\cos\frac{u}{2} + e^{u-2}\right] du \approx 1.430.$$

It follows that

$$\frac{2(\sqrt{1+d^2}-1)}{F_d} = \frac{\sqrt{5}-2}{F_d} \approx 0.356,$$

and

$$\frac{-1+\sqrt{1+\frac{4c^2}{N+1}}}{F_c} = \frac{\sqrt{5}-1}{F_c} \approx 0.865.$$

The above together with

$$\frac{2\sqrt{N(N+1)}\sin\frac{N\pi}{2(N+1)}}{\beta} = \frac{4\sqrt{3}\sin\frac{3\pi}{8}}{10} \approx 0.640$$

confirms condition (T_2) *of Theorem 2.*

Besides,

$$\begin{aligned}\liminf_{|\xi|\to+\infty}\frac{F(k,\xi)}{\xi} &= \liminf_{\xi\to+\infty}\frac{\frac{1}{6}\left(4\pi-3+e^{2\pi-2}\right)\xi - \frac{\pi}{3}(4\pi-3+e^{2\pi-2}) + \frac{e^{-2}}{6}(e^{2\pi-1}) + \frac{8}{3}}{\xi} \\ &= \frac{1}{6}\left(4\pi-3+e^{2\pi-2}\right) \\ &\approx 13.673 > \beta\end{aligned}$$

for each $k \in \mathbb{Z}(1, N)$. *We can see from Theorem 2 that the boundary value problem has at least two non-zero solutions for each* $\lambda \in \left(\frac{4\sqrt{3}\sin\frac{3\pi}{8}}{10}, \frac{\sqrt{5}-1}{3\int_0^2 \frac{1}{6}\left[(3-2u)\cos\frac{u}{2}+e^{u-2}\right]du} \right) \approx (0.640, 0.865)$.

5. Conclusions

A discrete Dirichlet boundary value problem involving the mean curvature operator is studied in this paper. Unlike the existing result in [33], we obtained different sufficient conditions of the existence of multiple solutions without assuming that the nonlinear term oscillates at infinity, as shown in Theorems 1 and 2. First, according to the research results of Bonanno in [22], we obtain at least three non-trivial solutions in Theorem 1. In addition, as a supplement to Theorem 1, we prove the existence of at least three positive solutions through the maximum principle. Note that inequality (3) plays an important role in the proof of Theorem 1. For the situation that inequality (3) is not satisfied, under another suitable assumption on the nonlinear term, we still can obtain the existence of at least two non-trivial solutions based on Theorem 2.1 in [23]. It seems that the method used in this paper can be adapted to discuss the existence of homoclinic solutions or periodic solutions of difference equations with ϕ_c-Laplacian. This will be left as our future work.

Author Contributions: All authors contributed equally and significantly in writing this paper. All authors have read and agreed to the published version of the manuscript.

Funding: This work is supported by the National Natural Science Foundation of China (Grant No. 11971126) and the Program for Changjiang Scholars and Innovative Research Team in University (Grant No. IRT 16R16).

Conflicts of Interest: The authors declare no conflict of interest.

References

1. Elaydi, S. *An Introduction to Difference Equations*; Springer Science & Business Media: Heidelberg, Germany, 2005.
2. Yu, J.; Zheng, B. Modeling Wolbachia infection in mosquito population via discrete dynamical models. *J. Differ. Equ. Appl.* **2019**, *25*, 1549–1567. [CrossRef]
3. Long, Y.; Wang, L. Global dynamics of a delayed two-patch discrete SIR disease model. *Commun. Nonlinear Sci. Numer. Simul.* **2020**, *83*, 105117. [CrossRef]
4. Agarwal, R.P. *Difference Equations and Inequalities: Theory, Methods, and Applications*; Marcel Dekker: New York, NY, USA, 1992.
5. Jiang, D.; Oregan, D.; Agarwal, R.P. A generalized upper and lower solution method for singular discrete boundary value problems for the one-dimensional *p*-Laplacian. *J. Appl. Anal.* **2005**, *11*, 35–47. [CrossRef]
6. Bereanu, C.; Mawhin, J. Existence and multiplicity results for nonlinear second order difference equations with Dirichlet boundary conditions. *Math. Bohem.* **2006**, *131*, 145–160. [CrossRef]
7. Long, Y.; Wang, S. Multiple solutions for nonlinear functional difference equations by the invariant sets of descending flow. *J. Differ. Equ. Appl.* **2019**, *25*, 1768–1789. [CrossRef]
8. Guo, Z.; Yu, J. Existence of periodic and subharmonic solutions for second-order superlinear difference equations. *Sci. China Ser. A Math.* **2003**, *46*, 506–515. [CrossRef]
9. Bereanu, C.; Jebelean, P.; Şerban, C. Periodic and Neumann problems for discrete $p(\cdot)$-Laplacian. *J. Math. Anal. Appl.* **2013**, *399*, 75–87. [CrossRef]
10. Shi, H. Periodic and subharmonic solutions for second-order nonlinear difference equations. *J. Appl. Math. Comput.* **2015**, *48*, 157–171. [CrossRef]
11. Shi, H.; Zhang, H. Existence of gap solitons in periodic discrete nonlinear Schrödinger equations. *J. Math. Anal. Appl.* **2010**, *361*, 411–419. [CrossRef]
12. Shi, H.; Zhang, Y. Existence of breathers for discrete nonlinear Schrödinger equations. *Appl. Math. Lett.* **2015**, *50*, 111–118. [CrossRef]
13. Lin, G.; Zhou, Z. Homoclinic solutions in non-periodic discrete ϕ-Laplacian equations with mixed nonlinearities. *Appl. Math. Lett.* **2017**, *64*, 15–20. [CrossRef]
14. Zhou, Z.; Yu, J. On the existence of homoclinic solutions of a class of discrete nonlinear periodic systems. *J. Differ. Equ.* **2010**, *249*, 1199–1212. [CrossRef]

15. Zhou, Z.; Ma, D. Multiplicity results of breathers for the discrete nonlinear Schrödinger equations with unbounded potentials. *Sci. China Math.* **2015**, *58*, 781–790. [CrossRef]
16. Nastasi, A.; Vetro, C. A note on homoclinic solutions of (p,q)-Laplacian difference equations. *J. Differ. Equ. Appl.* **2019**, *25*, 1–11. [CrossRef]
17. Galewski, M.; Smejda, J. On variational methods for nonlinear difference equations. *J. Comput. Appl. Math.* **2010**, *233*, 2985–2993. [CrossRef]
18. Jiang, L.; Zhou, Z. Three solutions to Dirichlet boundary value problems for p-Laplacian difference equations. *Adv. Differ. Equ.* **2007**, *2008*, 1–10. [CrossRef]
19. Long, Y.; Chen, J. Existence of multiple solutions to second-order discrete Neumann boundary value problems. *Appl. Math. Lett.* **2018**, *83*, 7–14. [CrossRef]
20. Bonanno, G.; Bella, B.D. A fourth-order boundary value problem for a Sturm–Liouville type equation. *Appl. Math. Comput.* **2010**, *217*, 3635–3640. [CrossRef]
21. Bonanno, G.; Jebelean, P.; Şerban, C. Superlinear discrete problems. *Appl. Mathe. Lett.* **2016**, *52*, 162–168. [CrossRef]
22. Bonanno, G.; Candito, P.; D'Aguí, G. Variational methods on finite dimensional Banach spaces and discrete problems. *Adv. Nonlinear Stud.* **2014**, *14*, 915–939. [CrossRef]
23. Bonanno, G.; D'Agui, G. Two non-zero solutions for elliptic Dirichlet problems. *Z. Anal. Anwend.* **2016**, *35*, 449–465. [CrossRef]
24. Agarwal, R.P.; Perera, K.; O'Regan, D. Multiple positive solutions of singular and nonsingular discrete problems via variational methods. *Nonlinear Anal. Theory Methods Appl.* **2004**, *58*, 69–73. [CrossRef]
25. Bonanno, G. A critical points theorem and nonlinear differential problems. *J. Glob. Optim.* **2004**, *28*, 249–258. [CrossRef]
26. Bonanno, G.; Candito, P. Nonlinear difference equations investigated via critical point methods. *Nonlinear Anal.* **2009**, *70*, 3180–3186. [CrossRef]
27. Nastasi, A.; Vetro, C.; Vetro, F. Positive solutions of discrete boundary value problems with the (p,q)-Laplacian operator. *Electron. J. Differ. Equ.* **2017**, *2017*, 1–12.
28. Bonanno, G.; Livrea, R.; Mawhin, J. Existence results for parametric boundary value problems involving the mean curvature operator. *Nonlinear Differ. Equ. Appl. NoDEA* **2015**, *22*, 411–426. [CrossRef]
29. Bonheure, D.; Habets, P.; Obersnel, F.; Omari, P. Classical and non-classical solutions of a prescribed curvature equation. *J. Differ. Equ.* **2007**, *243*, 208–237. [CrossRef]
30. Corsato, C.; Obersnel, F.; Omari, P.; Rivetti, S. Positive solutions of the Dirichlet problem for the prescribed mean curvature equation in Minkowski space. *J. Math. Anal. Appl.* **2013**, *405*, 227–239. [CrossRef]
31. Dai, G. Global bifurcation for problem with mean curvature operator on general domain. *Nonlinear Differ. Equ. Appl. NoDEA* **2017**, *3*, 1–10. [CrossRef]
32. Mawhin, J. Periodic solutions of second order nonlinear difference systems with ϕ-Laplacian: A variational approach. *Nonlinear Anal.* **2012**, *75*, 4672–4687. [CrossRef]
33. Zhou, Z.; Ling, J. Infinitely many positive solutions for a discrete two point nonlinear boundary value problem with ϕ_c-Laplacian. *Appl. Math. Lett.* **2019**, *91*, 28–34. [CrossRef]

Publisher's Note: MDPI stays neutral with regard to jurisdictional claims in published maps and institutional affiliations.

Article

On the Absolute Stable Difference Scheme for Third Order Delay Partial Differential Equations

Allaberen Ashyralyev [1,2,3,*], Evren Hincal [1] and Suleiman Ibrahim [1]

1 Department of Mathematics, Near East University, Lefkosa, Mersin 10, 99138 Turkey; evren.hincal@neu.edu.tr (E.H.); ibrahim.suleiman@neu.edu.tr (S.I.)
2 Department of Mathematics,Peoples' Friendship University of Russia (RUDN University), Moscow 117198, Russia
3 Institute of Mathematics and Mathematical Modeling, Almaty 050010, Kazakhstan
* Correspondence: allaberen.ashyralyev@neu.edu.tr(A.A)

Received: 16 April 2020; Accepted: 15 June 2020; Published: 19 June 2020

Abstract: The initial value problem for the third order delay differential equation in a Hilbert space with an unbounded operator is investigated. The absolute stable three-step difference scheme of a first order of accuracy is constructed and analyzed. This difference scheme is built on the Taylor's decomposition method on three and two points. The theorem on the stability of the presented difference scheme is proven. In practice, stability estimates for the solutions of three-step difference schemes for different types of delay partial differential equations are obtained. Finally, in order to ensure the coincidence between experimental and theoretical results and to clarify how efficient the proposed scheme is, some numerical experiments are tested.

Keywords: time delay; third order differential equations; difference scheme; stability

1. Introduction

Various problems in elasticity theory such as the problems of the longitudinal oscillations of a non-uniform viscoelastic rod, the problem of the longitudinal impact of a perfectly rigid body on a non-uniform finite-length viscoelastic rod with a variable cross-section, problems of wave propagation in a visco-elastic body, etc., lead to third order differential equations without the time delay term ([1–3]). Over the years, nonlocal and local boundary value problems have been of great interest due to their importance in the fields of engineering and science, especially in applied mathematics. Such problems have formed various research fields. Several nonlocal and local boundary value problems for differential equations have been investigated extensively in various works (for example, see [4–12] and the references given therein).

Differential equations having a delay term are used to model sociological, biological, as well as physical processes. They are used to model naturally occurring oscillation systems. A typical example of the occurrence of time delay can be seen in a sampled data control in control theory (see, for example, [13–17]). The presence of delay term in differential equations usually leads to difficulties in analyzing the differential equation. The boundedness and stability and the oscillation property of solutions for a third order delay ordinary differential and difference problems were widely studied (for example, see [18–25] and the references given therein).

Delay partial differential equations (DPDEs) arise in many applications such as control theory, climate models, medicine, biology, and much more (for example, see [26] and the references therein). The independent variables of partial differential equations having delay terms are time t together with one or more dimensional variable x, representing the position in space. It can also stand for the size of cells, relative DNA content, their level of mutation, as well as other parameters. The solutions of partial differential equations having delay terms may stand for voltage, temperature, or densities or concentrations of various particles, for instance chemicals, cells, animals, bacteria, and so on. Numerical methods for partial differential equations with delay terms usually lead to specific difficulties, which are usually not present in equations without delay terms. The theory and applications of parabolic and hyperbolic partial differential equations having a time delay term were studied by numerous authors (for example, see [13,27–34] and the references given therein). Recent publications on third order DPDEs are not many.

Several physical models lead to initial-boundary value problems for third order DPDEs (see, e.g., [1,3,8]). It is known that such types of problems can be replaced with the initial value problem for a third order delay differential equation:

$$\begin{cases} u_{ttt}(t) + Au_t(t) = bAu(t-w) + f(t),\ 0 < t < \infty, \\ u(t) = g(t), -w \le t \le 0 \end{cases} \tag{1}$$

in a Hilbert space H with unbounded operator A. Here, $b \in R^1$. Assume that $f(t)$ is a continuous function on $[0,\infty)$ and $f(t) \in D(A^{1/2})$, $g(t)$ is a twice continuously differentiable function on $[-w,0]$ and $g^{(k)}(t) \in D(A^{(3-k)/2})$ for $k = 0,1,2$.

Let us give the main theorem of paper [35].

Theorem 1. *The solution of Problem (1) satisfies the stability estimates:*

$$a_1 \le da_0 + \int_0^w \|f(s)\|_{D(A^{\frac{1}{2}})} ds, d = 2 + |b|\, w, \tag{2}$$

$$a_{n+1} \le da_n + \sum_{j=1}^{n+1} \int_{(j-1)\omega}^{jw} \|f(s)\|_{D(A^{\frac{1}{2}})} ds, n = 1,2,\cdots, \tag{3}$$

$$a_0 = \max\left\{ \max_{-w\le t\le 0} \|g_{tt}(t)\|_{D(A^{\frac{1}{2}})}, \max_{-w\le t\le 0} \|g_t(t)\|_{D(A)}, \max_{-w\le t\le 0} \|g(t)\|_{D(A^{\frac{3}{2}})} \right\},$$

$$a_n = \max\left\{ \max_{(n-1)w\le t\le nw} \|u_{tt}(t)\|_{D(A^{\frac{1}{2}})}, \max_{(n-1)w\le t\le nw} \|u_t(t)\|_{D(A)}, \max_{(n-1)w\le t\le nw} \|u(t)\|_{D(A^{\frac{3}{2}})} \right\}.$$

In practice, stability estimates for the solution of several problems for third order DPDEs were obtained.

Moreover, publications on the theory and applications of difference schemes (DSs) for third order DPDEs are not available. Thus, the construction and investigation of stable DSs for the approximate solutions of third order DPDEs is of great importance. Our aim in this paper is to construct the absolute stable three-step DS of the first order of accuracy of the third order DPDE for the approximate solution of the problem (1). We consider the uniform set of grid points:

$$[-w,\infty)_\tau = \{t_k : t_k = k\tau, -N \le k < \infty, N\tau = \omega\}$$

with step $\tau > 0$. Applying Taylor's decomposition method on three and two points (see [36,37]), we present the DS of the first order of accuracy:

$$\begin{cases} \frac{u_{k+2}-3u_{k+1}+3u_k-u_{k-1}}{\tau^3} + A\frac{u_{k+2}-u_{k+1}}{\tau} = bAu_{k-N} + f(t_k),\ k \ge 1, \\ u_k = g(t_k),\ -N+1 \le k \le 0, \\ (I+\tau^2 A)\frac{u_1-u_0}{\tau} = g'(0), (I+\tau^2 A)\frac{u_2-2u_1+u_0}{\tau^2} = g''(0), \\ (I+\tau^2 A)\frac{u_{mN+1}-u_{mN}}{\tau} = \frac{u_{mN}-u_{mN-1}}{\tau}, \\ (I+\tau^2 A)\frac{u_{mN+2}-2u_{mN+1}+u_{mN}}{\tau^2} = \frac{u_{mN}-2u_{mN-1}+u_{mN-2}}{\tau^2}, m = 1,2,... \end{cases} \tag{4}$$

for the approximate solution of Problem (1).

The organization of this paper is as follows. In Section 2, the main theorem on the stability of DS (4) is established. In Section 3, stability estimates of DSs for the approximate solution of three problems for third order DPDEs are obtained. Numerical results are provided for one= and two=dimensional third order DPDEs in Section 4. Finally, Section 5 gives the conclusion and our future plans.

2. Stability of DS

All over the present paper, assume that H is a Hilbert space and A is a self-adjoint positive definite operator $A \geq \delta I$ in H and $R = (I - i\tau A^{\frac{1}{2}})^{-1}$, $\widetilde{R} = (I + i\tau A^{\frac{1}{2}})^{-1}$.

Note that three-step DS (4) can obviously be rewritten as the system of single-step and two-step delay DS:

$$\begin{cases} \frac{u_{k+1}-u_k}{\tau} = v_k, k \geq 0, \\ \frac{v_{k+1}-2v_k+v_{k-1}}{\tau^2} + Av_{k+1} = p_k, p_k = bu_{k-N} + f(t_k), \ k \geq 1, \\ u_k = g(t_k), \ -N \leq k \leq 0, \\ (I+\tau^2 A)v_0 = g'(0), (I+\tau^2 A)\frac{v_1-v_0}{\tau} = g''(0), \\ (I+\tau^2 A)v_{mN} = \frac{u_{mN}-u_{mN-1}}{\tau}, \\ (I+\tau^2 A)\frac{v_{mN+1}-v_{mN}}{\tau} = \frac{u_{mN}-2u_{mN-1}+u_{mN-2}}{\tau^2}, m = 1, 2, ... \end{cases} \tag{5}$$

for the solution of DS (4). Applying DC(5), we can obtain the formula for the solution of DS (4). For this, we will consider two cases $1 \leq k \leq N$ and $mN+1 \leq k \leq (m+1)N$, $m = 1, 2, \cdots$, separately.

Let $1 \leq k \leq N$. Applying (5), we get the following DS:

$$\begin{cases} \frac{u_k-u_{k-1}}{\tau} = v_{k-1}, 1 \leq k \leq N, \\ \frac{v_{k+1}-2v_k+v_{k-1}}{\tau^2} + Av_{k+1} = p_k, p_k = bu_{k-N} + f(t_k), 1 \leq k \leq N-1, \\ u_k = g(t_k), \ -N \leq k \leq 0, \\ (I+\tau^2 A)v_0 = g'(0), (I+\tau^2 A)\frac{v_1-v_0}{\tau} = g''(0). \end{cases} \tag{6}$$

Therefore, we have that (see [38]):

$$u_0 = g(0), u_k = g(0) + \sum_{j=0}^{k-1} \tau v_j, \ 1 \leq k \leq N, \tag{7}$$

$$\begin{aligned} v_0 &= R\widetilde{R}g'(0), \ v_1 = R\widetilde{R}g'(0) + \tau R\widetilde{R}g''(0), \\ v_k &= \tfrac{1}{2}[R^{k-1} + \widetilde{R}^{k-1}]R\widetilde{R}g'(0) + \tfrac{1}{2i}A^{-\frac{1}{2}}R(R^k - \widetilde{R}^k)g''(0) \\ &\quad - \sum_{s=1}^{k-1} \tfrac{\tau}{2i}A^{-\frac{1}{2}}[R^{k-s} - \widetilde{R}^{k-s}]p_s \\ &= \tfrac{1}{2}[R^{k-1} + \widetilde{R}^{k-1}]R\widetilde{R}g'(0) + \tfrac{1}{2i}A^{-\frac{1}{2}}R(R^k - \widetilde{R}^k)g''(0) \\ &\quad + A^{-1}\left\{ \tfrac{1}{2}\sum_{s=2}^{k-1}[R^{k-s} + \widetilde{R}^{k-s}](p_{s-1} - p_s) + 2p_{k-1} - [R^{k-1} + \widetilde{R}^{k-1}]p_1 \right\}, 2 \leq k \leq N, \\ p_k &= bAg(t_{k-N}) + f(t_k), \ 1 \leq k \leq N-1. \end{aligned} \tag{8}$$

Applying Formulas (6)–(8), we obtain:

$$u_k=\begin{cases} g(0)+\tau R\widetilde{R}g'(0), k=0,\\ g(0)+\tau R\widetilde{R}g'(0), k=1,\\ g(0)+2\tau R\widetilde{R}g'(0)+\tau^2 R\widetilde{R}g''(0), k=2,\\ g(0)+\frac{1}{2i}A^{-\frac{1}{2}}\left(R^{k-2}-\widetilde{R}^{k-2}\right)R\widetilde{R}g'(0)-\frac{1}{2}A^{-1}R\left(2R\widetilde{R}-\left(R^{k-1}+\widetilde{R}^{k-1}\right)\right)g''(0)\\ +\sum\limits_{j=2}^{k-1}\tau\sum\limits_{s=1}^{j-1}\frac{\tau}{2i}A^{-\frac{1}{2}}[R^{j-s}-\widetilde{R}^{j-s}]p_s,\ 3\le k\le N. \end{cases}$$

By an interchange of the order of summation, we get:

$$u_k=\begin{cases} g(0)+\tau R\widetilde{R}g'(0), k=0,\\ g(0)+\tau R\widetilde{R}g'(0), k=1,\\ g(0)+2\tau R\widetilde{R}g'(0)+\tau^2 R\widetilde{R}g''(0), k=2,\\ g(0)+\frac{1}{2i}A^{-\frac{1}{2}}\left(R^{k-2}-\widetilde{R}^{k-2}\right)R\widetilde{R}g'(0)-\frac{1}{2}A^{-1}R\left(2R\widetilde{R}-\left(R^{k-1}+\widetilde{R}^{k-1}\right)\right)g''(0)\\ +A^{-1}\sum\limits_{s=1}^{k-2}\frac{\tau}{2}[2I-\left(R^{k-1-s}+\widetilde{R}^{k-1-s}\right)]p_s,\ 3\le k\le N \end{cases} \tag{9}$$

for the solution of DS (4).

Let $1+mN\le k\le(m+1)N,\ m=1,2,\cdots$. Applying (5), we can get the DS:

$$\begin{cases} \frac{u_k-u_{k-1}}{\tau}=v_{k-1}, mN+1\le k\le(m+1)N,\\ \frac{v_{k+1}-2v_k+v_{k-1}}{\tau^2}+Av_{k+1}=p_k, p_k=bu_{k-N}+f(t_k),\\ mN+1\le k\le(m+1)N-1,\\ u_{mN}\text{ is given, }(I+\tau^2A)v_{mN}=\frac{u_{mN}-u_{mN-1}}{\tau},\\ (I+\tau^2A)\frac{v_{mN+1}-v_{mN}}{\tau}=\frac{u_{mN}-2u_{mN-1}+u_{mN-2}}{\tau^2}. \end{cases} \tag{10}$$

Therefore, we have that (see [38]):

$$u_k=u_{mN}+\sum_{j=mN+2}^{k-1}\tau v_j,\ mN+1\le k\le(m+1)N, \tag{11}$$

$$v_{mN}=R\widetilde{R}\frac{u_{mN}-u_{mN-1}}{\tau},\ v_{mN+1}=R\widetilde{R}\frac{u_{mN}-u_{mN-1}}{\tau}+\tau R\widetilde{R}\frac{u_{mN}-2u_{mN-1}+u_{mN-2}}{\tau^2},$$

$$\begin{aligned} v_k&=\tfrac{1}{2}[R^{k-mN-1}+\widetilde{R}^{k-mN-1}]R\widetilde{R}\tfrac{u_{mN}-u_{mN-1}}{\tau}\\ &+\tfrac{1}{2i}A^{-\frac{1}{2}}R(R^{k-mN}-\widetilde{R}^{k-mN})\tfrac{u_{mN}-2u_{mN-1}+u_{mN-2}}{\tau^2}\\ &-\sum_{s=mN+1}^{k-1}\tfrac{\tau}{2i}A^{-\frac{1}{2}}[R^{k-s}-\widetilde{R}^{k-s}]p_s\\ &=\tfrac{1}{2}[R^{k-1}+\widetilde{R}^{k-1}]R\widetilde{R}\tfrac{u_{mN}-u_{mN-1}}{\tau}+\tfrac{1}{2i}A^{-\frac{1}{2}}R(R^k-\widetilde{R}^k)g''(0)\\ &+A^{-1}\left\{\tfrac{1}{2}\sum_{s=mN+2}^{k-1}[R^{k-s}+\widetilde{R}^{k-s}](p_{s-1}-p_s)\right.\\ &\left.+2p_{k-mN-1}-[R^{k-mN-1}+\widetilde{R}^{k-mN-1}]p_{mN+1}\right\}, mN+2\le k\le(m+1)N. \end{aligned} \tag{12}$$

Applying Formulas (10)–(12), we can obtain:

$$u_k = \begin{cases} u_{mN} + \tau R\widetilde{R}\frac{u_{mN}-u_{mN-1}}{\tau}, k = mN+1, \\ u_{mN} + 2\tau R\widetilde{R}\frac{u_{mN}-u_{mN-1}}{\tau} + \tau^2 R\widetilde{R}\frac{u_{mN}-2u_{mN-1}+u_{mN-2}}{\tau^2}, k = mN+2, \\ u_{mN} + \frac{1}{2i}A^{-\frac{1}{2}}\left(R^{k-mN-2} - \widetilde{R}^{k-mN-2}\right)R\widetilde{R}\frac{u_{mN}-u_{mN-1}}{\tau} \\ -\frac{1}{2}A^{-1}R\left(2R\widetilde{R} - \left(R^{k-mN-1} + \widetilde{R}^{k-mN-1}\right)\right)\frac{u_{mN}-2u_{mN-1}+u_{mN-2}}{\tau^2} \\ + \sum\limits_{j=mN+2}^{k-1} \tau \sum\limits_{s=mN+1}^{j-1} \frac{\tau}{2i}A^{-\frac{1}{2}}[R^{j-s} - \widetilde{R}^{j-s}]p_s,\ mN+3 \le k \le (m+1)N. \end{cases}$$

By an interchange of the order of summation, we get the solution of DS (4):

$$u_k = \begin{cases} u_{mN} + \tau R\widetilde{R}\frac{u_{mN}-u_{mN-1}}{\tau}, k = mN+1, \\ u_{mN} + 2\tau R\widetilde{R}\frac{u_{mN}-u_{mN-1}}{\tau} + \tau^2 R\widetilde{R}\frac{u_{mN}-2u_{mN-1}+u_{mN-2}}{\tau^2}, k = mN+2, \\ u_{mN} + \frac{1}{2i}A^{-\frac{1}{2}}\left(R^{k-mN-2} - \widetilde{R}^{k-mN-2}\right)R\widetilde{R}\frac{u_{mN}-u_{mN-1}}{\tau} \\ -\frac{1}{2}A^{-1}R\left(2R\widetilde{R} - \left(R^{k-mN-1} + \widetilde{R}^{k-mN-1}\right)\right)\frac{u_{mN}-2u_{mN-1}+u_{mN-2}}{\tau^2} \\ +A^{-1}\sum\limits_{s=mN+1}^{k-2} \frac{\tau}{2}[2I - \left(R^{k-1-s} + \widetilde{R}^{k-1-s}\right)]p_s,\ mN+3 \le k \le (m+1)N. \end{cases} \quad (13)$$

The following lemma will be needed in the sequel.

Lemma 1. *The following estimates are fulfilled:*

$$\|R\|_{H\to H},\ \|\widetilde{R}\|_{H\to H} \le 1, \quad (14)$$

$$\|R\widetilde{R}^{-1}\|_{H\to H},\ \|\widetilde{R}R^{-1}\|_{H\to H} \le 1, \quad (15)$$

$$\|\tau A^{\frac{1}{2}}R\|_{H\to H},\ \|\tau A^{\frac{1}{2}}\widetilde{R}\|_{H\to H} \le 1. \quad (16)$$

The proof of the estimates (14)–(16) is based on the spectral theory of a self-adjoint operator in a Hilbert space [39].

Now, let us study the stability of DS (4).

Theorem 2. *The solution of DS (4) satisfies the following stability estimates:*

$$b_1 \le (2 + \tau|b|(N-2))b_0 + \tau\sum_{s=1}^{N-2}\|A^{\frac{1}{2}}f(t_s)\|_H, \quad (17)$$

$$b_{m+1} \le (2 + \tau|b|(N-2))b_m + \tau\sum_{s=mN+1}^{(m+1)N}\|A^{\frac{1}{2}}f(t_s)\|_H, \quad (18)$$

$$b_0 = \max\left\{\max_{-N\le k\le 0}\|A^{\frac{1}{2}}g_{tt}(t_k)\|_H,\ \max_{-N\le k\le 0}\|Ag_t(t_k)\|_H,\ \max_{-N\le k\le 0}\|A^{\frac{3}{2}}g(t_k)\|_H\right\},$$

$$b_m = \max\left\{\max_{(m-1)N\le k\le mN-2}\|A^{\frac{1}{2}}\frac{u_{k+2}-2u_{k+1}+u_k}{\tau^2}\|_H,\right.$$

$$\left.\max_{(m-1)N+1\le k\le mN}\|A\frac{u_k-u_{k-1}}{\tau}\|_H,\ \frac{1}{2}\max_{(m-1)N\le k\le mN}\|A^{\frac{3}{2}}u_k\|_H\right\},m = 1,2,\cdots.$$

Proof. Let us estimate b_1. Using Formula (9) and the estimates (14)–(16), we get that:

$$\|A^{\frac{3}{2}}u_1\|_H \le \|A^{\frac{3}{2}}g(0)\|_H + \|\tau A^{\frac{1}{2}}R\|_{H\to H}\|\widetilde{R}\|_{H\to H}\|Ag_t(0)\|_H \le 2b_0,$$

$$\|A^{\frac{3}{2}}u_2\|_H \le \|A^{\frac{3}{2}}g(0)\|_H + 2\|\tau A^{\frac{1}{2}}R\|_{H\to H}\|\widetilde{R}\|_{H\to H}\|Ag_t(0)\|_H$$
$$+\|\tau A^{\frac{1}{2}}R\|_{H\to H}\|\tau A^{\frac{1}{2}}\widetilde{R}\|_{H\to H}\|A^{\frac{1}{2}}g_{tt}(0)\|_H \le 4b_0,$$
$$\|A^{\frac{3}{2}}u_k\|_H \le \|A^{\frac{3}{2}}g(0)\|_H + \frac{1}{2}[\|R\|^{k-2}_{H\to H} + \|\widetilde{R}\|^{k-2}_{H\to H}]\|R\widetilde{R}\|_{H\to H}\|Ag_t(0)\|_H$$
$$+\frac{1}{2}\|R\|_{H\to H}[2\|R\widetilde{R}\|_{H\to H} + \|R\|^{k-1}_{H\to H} + \|\widetilde{R}\|^{k-1}_{H\to H}]\|A^{\frac{1}{2}}g_{tt}(0)\|_H$$
$$+\frac{\tau}{2}|b|\sum_{s=1}^{k-2}[2 + \|R\|^{k-1-s}_{H\to H} + \|\widetilde{R}\|^{k-1-s}_{H\to H}]\|A^{\frac{3}{2}}g(t_{s-N})\|_H$$
$$+\frac{\tau}{2}\sum_{s=1}^{k-2}[2 + \|R\|^{k-1-s}_{H\to H} + \|\widetilde{R}\|^{k-1-s}_{H\to H}]\|A^{\frac{1}{2}}f(t_s)\|_H \le \|A^{\frac{3}{2}}g(0)\|_H + \|Ag_t(0)\|_H$$
$$+2\|A^{\frac{1}{2}}g_{tt}(0)\|_H + 2\tau|b|(N-2)\max_{-N\le k\le 0}\|A^{\frac{3}{2}}g(t_k)\|_H + 2\tau\sum_{s=1}^{k-2}\|A^{\frac{1}{2}}f(t_s)\|_H$$
$$\le 2\Big(2 + \tau|b|(N-2)\Big)b_0 + 2\tau\sum_{s=1}^{N-2}\|A^{\frac{1}{2}}f(t_s)\|_H,\ 3\le k\le N.$$

From that and $u_0 = g(0)$, it follows that:

$$\frac{1}{2}\max_{0\le k\le N}\|A^{\frac{3}{2}}u_k\|_H$$
$$\le (2 + \tau|b|(N-2))b_0 + \tau\sum_{s=1}^{N-2}\|A^{\frac{1}{2}}f(t_s)\|_H \tag{19}$$

for the solution of DS (4). Applying Formulas (6)–(8), we can write:

$$\frac{u_1 - u_0}{\tau} = v_0 = R\widetilde{R}g_t(0),$$
$$\frac{u_2 - u_1}{\tau} = v_1 = R\widetilde{R}g_t(0) + \tau R\widetilde{R}g_{tt}(0),$$
$$\frac{u_k - u_{k-1}}{\tau} = v_{k-1} = \frac{1}{2}[R^{k-2} + \widetilde{R}^{k-2}]R\widetilde{R}g_t(0) + \frac{1}{2i}R(R^{k-1} - \widetilde{R}^{k-1})A^{-\frac{1}{2}}g_{tt}(0)$$
$$-\frac{\tau}{2i}\sum_{s=1}^{k-2}[R^{k-1-s} - \widetilde{R}^{k-1-s}]bA^{\frac{1}{2}}g(t_{s-N}) - \frac{\tau}{2i}\sum_{s=1}^{k-2}[R^{k-1-s} - \widetilde{R}^{k-1-s}]A^{-\frac{1}{2}}f(t_s),\ 3\le k\le N.$$

Using this formula and the estimates (14)–(16), we obtain:

$$\|A\frac{u_1 - u_0}{\tau}\|_H \le \|R\widetilde{R}\|_{H\to H}\|g_t(0)\|_H \le b_0,$$
$$\|A\frac{u_2 - u_1}{\tau}\|_H \le \|R\widetilde{R}\|_{H\to H}\|Ag_t(0)\|_H + \|\tau A^{\frac{1}{2}}R\|_{H\to H}\|\widetilde{R}\|_{H\to H}\|A^{\frac{1}{2}}g_{tt}(0)\|_H \le 2b_0,$$
$$\|A\frac{u_k - u_{k-1}}{\tau}\|_H \le \frac{1}{2}[\|R\|^{k-2}_{H\to H} + \|\widetilde{R}\|^{k-2}_{H\to H}]\|R\widetilde{R}\|_{H\to H}\|g_t(0)\|_H$$
$$+\frac{1}{2}[\|R\|^{k-1}_{H\to H} + \|\widetilde{R}\|^{k-1}_{H\to H}]\|R\|_{H\to H}\|A^{\frac{1}{2}}g_{tt}(0)\|_H + \frac{\tau}{2}|b|\sum_{s=1}^{N-2}[\|R\|^{k-1-s}_{H\to H} + \|\widetilde{R}\|^{k-1-s}_{H\to H}]\|A^{\frac{3}{2}}g(t_{s-N})\|_H$$
$$+\frac{\tau}{2}\sum_{s=1}^{N-2}[\|R\|^{k-1-s}_{H\to H} + \|\widetilde{R}\|^{k-1-s}_{H\to H}]\|A^{\frac{1}{2}}f(t_s)\|_H$$

$$\leq (2+\tau|b|(N-2))b_0+\tau\sum_{s=1}^{N-2}\|A^{\frac{1}{2}}f(t_s)\|_H, 3\leq k\leq N.$$

Combining these estimates, we obtain:

$$\max_{1\leq k\leq N}\left\|A\frac{u_k-u_{k-1}}{\tau}\right\|_H\leq (2+\tau|b|(N-2))b_0+\tau\sum_{s=1}^{N-2}\|A^{\frac{1}{2}}f(t_s)\|_H \tag{20}$$

for the solution of DS (4). Applying Formulas (6)–(8), we can write:

$$\frac{u_2-2u_1+u_0}{\tau^2}=\tau^{-1}\left(v_1-v_0\right)=\widetilde{R}Rg_{tt}(0),$$

$$\frac{u_3-2u_2+u_1}{\tau^2}=\tau^{-1}\left(v_2-v_1\right)$$

$$=-\tau A(R\widetilde{R})^2g_t(0)+\left(I-\tau^2A\right)\left(\widetilde{R}R\right)^2g_{tt}(0)+\tau R\widetilde{R}p_1$$

$$=-\tau A(R\widetilde{R})^2g_t(0)+\left(I-\tau^2A\right)\left(\widetilde{R}R\right)^2g_{tt}(0)+\tau R\widetilde{R}[bAg(t_{1-N})+f(t_1)],$$

$$\frac{u_{k+2}-2u_{k+1}+u_k}{\tau^2}=\tau^{-1}\left(v_{k+1}-v_k\right)$$

$$=\tau^{-1}\left\{\frac{1}{2}[R^k+\widetilde{R}^k]R\widetilde{R}g_t(0)+\frac{1}{2i}A^{-\frac{1}{2}}[R^{k+1}-\widetilde{R}^{k+1}]Rg_{tt}(0)\right.$$

$$-\sum_{s=1}^{k}\frac{\tau}{2i}A^{-\frac{1}{2}}[R^{k+1-s}-\widetilde{R}^{k+1-s}]p_s-\frac{1}{2}[R^{k-1}+\widetilde{R}^{k-1}]R\widetilde{R}g_t(0)$$

$$\left.-\frac{1}{2i}A^{-\frac{1}{2}}[R^k-\widetilde{R}^k]Rg_{tt}(0)+\sum_{s=1}^{k}\frac{\tau}{2i}A^{-\frac{1}{2}}[R^{k-s}-\widetilde{R}^{k-s}]p_s\right\}$$

$$=\tau^{-1}\left\{\frac{1}{2}[R^k+\widetilde{R}^k-R^{k-1}-\widetilde{R}^{k-1}]R\widetilde{R}g_t(0)+\frac{1}{2i}A^{-1}[R^{k+1}-\widetilde{R}^{k+1}-R^k+\widetilde{R}^k]Rg_{tt}(0)\right.$$

$$\left.-\frac{\tau}{2i}A^{-\frac{1}{2}}[R-\widetilde{R}]p_k+\sum_{s=1}^{k-1}\frac{\tau}{2i}A^{-\frac{1}{2}}[R^{k-s}-\widetilde{R}^{k-s}-R^{k+1-s}+\widetilde{R}^{k+1-s}]p_s\right\}$$

$$=-\frac{i}{2}[R^k-\widetilde{R}^k]R\widetilde{R}A^{\frac{1}{2}}g_t(0)+\frac{1}{2}R[R^{k+1}+\widetilde{R}^{k+1}]g_{tt}(0)-\tau R\widetilde{R}Abg(t_{k-N})-\tau R\widetilde{R}f(t_k)$$

$$-\frac{\tau}{2}\sum_{s=1}^{k-1}[R^{k-s+1}+\widetilde{R}^{k-s+1}]Abg(t_{s-N})-\frac{\tau}{2}\sum_{s=1}^{k-1}[R^{k-s+1}+\widetilde{R}^{k-s+1}]f(t_s), 2\leq k\leq N-2.$$

Using this formula and the estimates (14)–(16), we obtain:

$$\left\|A^{\frac{1}{2}}\frac{u_2-2u_1+u_0}{\tau^2}\right\|_H\leq\|R\widetilde{R}\|_{H\to H}\|A^{\frac{1}{2}}g_{tt}(0)\|_H\leq b_0,$$

$$\left\|A^{\frac{1}{2}}\frac{u_3-2u_2+u_1}{\tau^2}\right\|_H\leq\|\tau A^{\frac{1}{2}}R\|_{H\to H}\|R\|_{H\to H}\|\widetilde{R}\|_{H\to H}^2\|Ag_t(0)\|_H+\|R\widetilde{R}\|_{H\to H}\|A^{\frac{1}{2}}g_{tt}(0)\|_H$$

$$+\tau|b|\|R\widetilde{R}\|_{H\to H}\|A^{\frac{3}{2}}g(t_{1-N})\|_H+\tau\|R\widetilde{R}\|_{H\to H}\|A^{\frac{3}{2}}f(t_1)\|_H$$

$$\leq (2+\tau|b|)b_0+\tau\|A^{\frac{1}{2}}f(t_1)\|_H,$$

$$\left\|A^{\frac{1}{2}}\frac{u_{k+2}-2u_{k+1}+u_k}{\tau^2}\right\|_H\leq\frac{1}{2}[\|R\|_{H\to H}^k+\|\widetilde{R}\|_{H\to H}^k]\|R\widetilde{R}\|_{H\to H}\|Ag_t(0)\|_H$$

$$+\frac{1}{2}\|R\|_{H\to H}[\|R\|_{H\to H}^{k+1}+\|\widetilde{R}\|_{H\to H}^{k+1}]\|A^{\frac{1}{2}}g_{tt}(0)\|_H+\frac{\tau}{2}|b|\sum_{s=1}^{k-1}[\|R\|_{H\to H}^{k-s+1}+\|\widetilde{R}\|_{H\to H}^{k-s+1}]\|A^{\frac{3}{2}}g(t_{s-N})\|_H$$

$$+\frac{\tau}{2}\sum_{s=1}^{k-1}[\|R\|_{H\to H}^{k-s+1}+\|\widetilde{R}\|_{H\to H}^{k-s+1}]\|A^{\frac{1}{2}}f(t_s)\|_H+\tau|b|\|R\widetilde{R}\|_{H\to H}\|A^{\frac{3}{2}}g(t_{k-N})\|_H$$

$$+\tau\|R\widetilde{R}\|_{H\to H}\|A^{\frac{1}{2}}f(t_k)\|_H$$

$$\le(2+\tau|b|(N-2))b_0+\tau\sum_{s=1}^{N-2}\|A^{\frac{1}{2}}f(t_s)\|_H, 2\le k\le N-2.$$

Combining these estimates, we can get:

$$\max_{0\le k\le N-2}\left\|A^{\frac{1}{2}}\frac{u_{k+2}-2u_{k+1}+u_k}{\tau^2}\right\|_H$$

$$\le(2+\tau|b|(N-2))b_0+\tau\sum_{s=1}^{N-2}\|A^{\frac{1}{2}}f(t_s)\|_H \tag{21}$$

for the solution of DS (4). Estimate (17) follows from (19)–(21).

Now, let us estimate b_{m+1}. Using Formula (13) and the estimates (14)–(16), we can obtain:

$$\|A^{\frac{3}{2}}u_{mN+1}\|_H\le\|A^{\frac{3}{2}}u_{mN}\|_H+\|\tau A^{\frac{1}{2}}R\|_{H\to H}\|\widetilde{R}\|_{H\to H}\|A\frac{u_{mN}-u_{mN-1}}{\tau}\|_H\le 2b_m,$$

$$\|A^{\frac{3}{2}}u_{mN+2}\|_H\le\|A^{\frac{3}{2}}u_{mN}\|_H+2\|\tau A^{\frac{1}{2}}R\|_{H\to H}\|\widetilde{R}\|_{H\to H}\|A\frac{u_{mN}-u_{mN-1}}{\tau}\|_H$$

$$+\|\tau A^{\frac{1}{2}}R\|_{H\to H}\|\tau A^{\frac{1}{2}}\widetilde{R}\|_{H\to H}\|A^{\frac{1}{2}}\frac{u_{mN}-2u_{mN-1}+u_{mN-2}}{\tau^2}\|_H\le 4b_m,$$

$$\|A^{\frac{3}{2}}u_k\|_H\le\|A^{\frac{3}{2}}u_{mN}\|_H+\frac{1}{2}[\|R\|_{H\to H}^{k-mN-2}+\|\widetilde{R}\|_{H\to H}^{k-mN-2}]\|R\widetilde{R}\|_{H\to H}\|A\frac{u_{mN}-u_{mN-1}}{\tau}\|_H$$

$$+\frac{1}{2}\|R\|_{H\to H}[2\|R\widetilde{R}\|_{H\to H}+\|R\|_{H\to H}^{k-mN-1}+\|\widetilde{R}\|_{H\to H}^{k-mN-1}]\|A^{\frac{1}{2}}\frac{u_{mN}-2u_{mN-1}+u_{mN-2}}{\tau^2}\|_H$$

$$+\frac{\tau}{2}|b|\sum_{s=mN+1}^{k-2}[2+\|R\|_{H\to H}^{k-1-s}+\|\widetilde{R}\|_{H\to H}^{k-1-s}]\|A^{\frac{3}{2}}u(t_{s-N})\|_H$$

$$+\frac{\tau}{2}\sum_{s=mN+1}^{k-2}[2+\|R\|_{H\to H}^{k-1-s}+\|\widetilde{R}\|_{H\to H}^{k-1-s}]\|A^{\frac{1}{2}}f(t_s)\|_H$$

$$\le 2\Big(2+\tau|b|(N-2)\Big)b_m+2\tau\sum_{s=mN+1}^{(m+1)N-2}\|A^{\frac{1}{2}}f(t_s)\|_H,\ mN+3\le k\le(m+1)N.$$

Combining these estimates, we obtain:

$$\frac{1}{2}\max_{mN+1\le k\le(m+1)N}\|A^{\frac{3}{2}}u_k\|_H$$

$$\le(2+\tau|b|(N-2))b_m+\tau\sum_{s=mN+1}^{(m+1)N-2}\|A^{\frac{1}{2}}f(t_s)\|_H \tag{22}$$

for the solution of DS (4). Applying Formulas (10)–(12), we can write:

$$\frac{u_{mN+1}-u_{mN}}{\tau}=v_{mN}=R\widetilde{R}\frac{u_{mN}-u_{mN-1}}{\tau},$$

$$\frac{u_{mN+2}-u_{mN+1}}{\tau}=v_{mN+1}=R\widetilde{R}\frac{u_{mN}-u_{mN-1}}{\tau}+\tau R\widetilde{R}\frac{u_{mN}-2u_{mN-1}+u_{mN-2}}{\tau^2},$$

$$\frac{u_k-u_{k-1}}{\tau}=v_{k-1}=\frac{1}{2}[R^{k-mN-2}+\widetilde{R}^{k-mN-2}]R\widetilde{R}\frac{u_{mN}-u_{mN-1}}{\tau}$$
$$+\frac{1}{2i}R(R^{k-mN-1}-\widetilde{R}^{k-mN-1})A^{-\frac{1}{2}}\frac{u_{mN}-2u_{mN-1}+u_{mN-2}}{\tau^2}$$
$$-\frac{\tau}{2i}\sum_{s=mN+1}^{k-2}[R^{k-1-s}-\widetilde{R}^{k-1-s}]bA^{\frac{1}{2}}u(t_{s-N})$$
$$-\frac{\tau}{2i}\sum_{s=mN+1}^{k-2}[R^{k-1-s}-\widetilde{R}^{k-1-s}]A^{-\frac{1}{2}}f(t_s),\ mN+3\le k\le(m+1)N.$$

Using this formula and the estimates (14)–(16), we can obtain:

$$\|A\frac{u_{mN+1}-u_{mN}}{\tau}\|_H\le\|R\widetilde{R}\|_{H\to H}\|A\frac{u_{mN}-u_{mN-1}}{\tau}\|_H\le b_m,$$

$$\|A\frac{u_{mN+2}-u_{mN+1}}{\tau}\|_H\le\|R\widetilde{R}\|_{H\to H}\|A\frac{u_{mN}-u_{mN-1}}{\tau}\|_H$$
$$+\|\tau A^{\frac{1}{2}}R\|_{H\to H}\|\widetilde{R}\|_{H\to H}\|A^{\frac{1}{2}}\frac{u_{mN}-2u_{mN-1}+u_{mN-2}}{\tau^2}\|_H\le 2b_m,$$

$$\|A\frac{u_k-u_{k-1}}{\tau}\|_H\le\frac{1}{2}[\|R\|_{H\to H}^{k-mN-2}+\|\widetilde{R}\|_{H\to H}^{k-mN-2}]\|R\widetilde{R}\|_{H\to H}\|A\frac{u_{mN}-u_{mN-1}}{\tau}\|_H$$
$$+\frac{1}{2}[\|R\|_{H\to H}^{k-1}+\|\widetilde{R}\|_{H\to H}^{k-1}]\|R\|_{H\to H}\|A^{\frac{1}{2}}\frac{u_{mN}-2u_{mN-1}+u_{mN-2}}{\tau^2}\|_H$$
$$+\frac{\tau}{2}|b|\sum_{s=mN+1}^{(m+1)N-2}[\|R\|_{H\to H}^{k-1-s}+\|\widetilde{R}\|_{H\to H}^{k-1-s}]\|A^{\frac{3}{2}}u(t_{s-N})\|_H$$
$$+\frac{\tau}{2}\sum_{s=mN+1}^{(m+1)N-2}[\|R\|_{H\to H}^{k-1-s}+\|\widetilde{R}\|_{H\to H}^{k-1-s}]\|A^{\frac{1}{2}}f(t_s)\|_H$$
$$\le(2+\tau|b|(N-2))b_m+\tau\sum_{s=mN+1}^{(m+1)N-2}\|A^{\frac{1}{2}}f(t_s)\|_H,\ mN+3\le k\le(m+1)N.$$

Combining these estimates, we obtain:

$$\max_{mN+1\le k\le(m+1)N}\left\|A\frac{u_k-u_{k-1}}{\tau}\right\|_H\le(2+\tau|b|(N-2))b_m+\tau\sum_{s=mN+1}^{(m+1)N-2}\|A^{\frac{1}{2}}f(t_s)\|_H \quad (23)$$

for the solution of DS (4). Applying Formulas (10)–(12), we can write:

$$\frac{u_{mN+2}-2u_{mN+1}+u_{mN}}{\tau^2}=\tau^{-1}\left(v_{mN+1}-v_{mN}\right)=\widetilde{R}R\frac{u_{mN}-2u_{mN-1}+u_{mN-2}}{\tau^2},$$

$$\frac{u_{mN+3}-2u_{mN+2}+u_{mN+1}}{\tau^2}=\tau^{-1}\left(v_{mN+2}-v_{mN+1}\right)$$
$$=-\tau A(R\widetilde{R})^2\frac{u_{mN}-u_{mN-1}}{\tau}+\left(I-\tau^2A\right)\left(\widetilde{R}R\right)^2\frac{u_{mN}-2u_{mN-1}+u_{mN-2}}{\tau^2}+\tau R\widetilde{R}p_{mN+1}$$
$$=-\tau A(R\widetilde{R})^2\frac{u_{mN}-u_{mN-1}}{\tau}+\left(I-\tau^2A\right)\left(\widetilde{R}R\right)^2\frac{u_{mN}-2u_{mN-1}+u_{mN-2}}{\tau^2}$$

$$\frac{u_{k+2}-2u_{k+1}+u_k}{\tau^2}=\tau^{-1}\left(v_{k+1}-v_k\right)$$

$$+\tau R\widetilde{R}[bAu(t_{1-N})+f(t_{mN+1})],$$

$$=\tau^{-1}\left\{\frac{1}{2}[R^{k-mN}+\widetilde{R}^{k-mN}]R\widetilde{R}\frac{u_{mN}-u_{mN-1}}{\tau}\right.$$

$$+\frac{1}{2i}A^{-\frac{1}{2}}[R^{k-Mn+1}-\widetilde{R}^{k-Mn+1}]R\frac{u_{mN}-2u_{mN-1}+u_{mN-2}}{\tau^2}$$

$$-\sum_{s=Mn+1}^{k}\frac{\tau}{2i}A^{-\frac{1}{2}}[R^{k+1-s}-\widetilde{R}^{k+1-s}]p_s-\frac{1}{2}[R^{k-Mn-1}+\widetilde{R}^{k-Mn-1}]R\widetilde{R}\frac{u_{mN}-u_{mN-1}}{\tau}$$

$$-\frac{1}{2i}A^{-\frac{1}{2}}[R^{k-mN}-\widetilde{R}^{k-mN}]R\frac{u_{mN}-2u_{mN-1}+u_{mN-2}}{\tau^2}$$

$$\left.+\sum_{s=mN+1}^{k}\frac{\tau}{2i}A^{-\frac{1}{2}}[R^{k-s}-\widetilde{R}^{k-s}]p_s\right\}$$

$$=\tau^{-1}\left\{\frac{1}{2}[R^{k-mN}+\widetilde{R}^{k-mN}-R^{k-mN-1}-\widetilde{R}^{k-mN-1}]R\widetilde{R}\frac{u_{mN}-u_{mN-1}}{\tau}\right.$$

$$+\frac{1}{2i}A^{-1}[R^{k+1}-\widetilde{R}^{k+1}-R^{k}+\widetilde{R}^{k}]R\frac{u_{mN}-2u_{mN-1}+u_{mN-2}}{\tau^2}$$

$$\left.-\frac{\tau}{2i}A^{-\frac{1}{2}}[R-\widetilde{R}]p_{k-mN}+\sum_{s=mN+1}^{k-1}\frac{\tau}{2i}A^{-\frac{1}{2}}[R^{k-s}-\widetilde{R}^{k-s}-R^{k+1-s}+\widetilde{R}^{k+1-s}]p_s\right\}$$

$$=-\frac{i}{2}[R^{k-mN}-\widetilde{R}^{k-mN}]R\widetilde{R}A^{\frac{1}{2}}\frac{u_{mN}-u_{mN-1}}{\tau}$$

$$+\frac{1}{2}R[R^{k+1}+\widetilde{R}^{k+1}]\frac{u_{mN}-2u_{mN-1}+u_{mN-2}}{\tau^2}$$

$$-\tau R\widetilde{R}Abu(t_{k-N})-\tau R\widetilde{R}f(t_{k-mN})-\frac{\tau}{2}\sum_{s=mN+1}^{k-1}[R^{k-s+1}+\widetilde{R}^{k-s+1}]Abu(t_{s-N})$$

$$-\frac{\tau}{2}\sum_{s=mN+1}^{k-1}[R^{k-s+1}+\widetilde{R}^{k-s+1}]f(t_s),\ mN+2\le k\le(m+1)N-2.$$

Using this formula and the estimates (14)–(16), we obtain that:

$$\left\|A^{\frac{1}{2}}\frac{u_{mN+2}-2u_{mN+1}+u_{mN}}{\tau^2}\right\|_H\le\|R\widetilde{R}\|_{H\to H}\|A^{\frac{1}{2}}\frac{u_{mN}-2u_{mN-1}+u_{mN-2}}{\tau^2}\|_H\le b_m,$$

$$\left\|A^{\frac{1}{2}}\frac{u_{mN+3}-2u_{mN+2}+u_{mN+1}}{\tau^2}\right\|_H$$

$$\le\|\tau A^{\frac{1}{2}}R\|_{H\to H}\|R\|_{H\to H}\|\widetilde{R}\|_{H\to H}^2\|A\frac{u_{mN}-u_{mN-1}}{\tau}\|_H$$

$$+\|R\widetilde{R}\|_{H\to H}\|A^{\frac{1}{2}}\frac{u_{mN}-2u_{mN-1}+u_{mN-2}}{\tau^2}\|_H$$

$$+\tau|b|\|R\widetilde{R}\|_{H\to H}\|A^{\frac{3}{2}}u(t_{1-N})\|_H+\tau\|R\widetilde{R}\|_{H\to H}\|A^{\frac{3}{2}}f(t_{mN+1})\|_H$$

$$\le(2+\tau|b|)b_m+\tau\|A^{\frac{1}{2}}f(t_{mN+1})\|_H,$$

$$\left\|A^{\frac{1}{2}}\frac{u_{k+2}-2u_{k+1}+u_k}{\tau^2}\right\|_H$$

$$\le\frac{1}{2}[\|R\|_{H\to H}^{k-mN}+\|\widetilde{R}\|_{H\to H}^{k-mN}]\|R\widetilde{R}\|_{H\to H}\|A\frac{u_{mN}-u_{mN-1}}{\tau}\|_H$$

$$+\frac{1}{2}\|R\|_{H\to H}[\|R\|_{H\to H}^{k+1}+\|\widetilde{R}\|_{H\to H}^{k+1}]\|A^{\frac{1}{2}}\frac{u_{mN}-2u_{mN-1}+u_{mN-2}}{\tau^2}\|_H$$

$$+\frac{\tau}{2}|b|\sum_{s=mN+1}^{k-1}[\|R\|_{H\to H}^{k-s+1}+\|\widetilde{R}\|_{H\to H}^{k-s+1}]\|A^{\frac{3}{2}}u(t_{s-N})\|_H$$

$$+\frac{\tau}{2}\sum_{s=mN+1}^{k-1}[\|R\|_{H\to H}^{k-s+1}+\|\widetilde{R}\|_{H\to H}^{k-s+1}]\|A^{\frac{1}{2}}f(t_s)\|_H$$

$$+\tau|b|\|R\widetilde{R}\|_{H\to H}\|A^{\frac{3}{2}}u(t_{k-(m+1)N})\|_H+\tau\|R\widetilde{R}\|_{H\to H}\|A^{\frac{1}{2}}f(t_{k-mN})\|_H$$

$$\leq(2+\tau|b|(N-2))b_m+\tau\sum_{s=mN+1}^{(m+1)N-2}\|A^{\frac{1}{2}}f(t_s)\|_H,\ mN+2\leq k\leq(m+1)N-2.$$

Combining these estimates, we obtain:

$$\max_{mN\leq k\leq(m+1)N-2}\left\|A^{\frac{1}{2}}\frac{u_{k+2}-2u_{k+1}+u_k}{\tau^2}\right\|_H$$

$$\leq(2+\tau|b|(N-2))b_m+\tau\sum_{s=mN+1}^{(m+1)N-2}\|A^{\frac{1}{2}}f(t_s)\|_H \tag{24}$$

for the solution of DS (4). Estimate (18) follows from (22)–(24). Theorem 2 is proven.

Note that applying Theorem 2, we can obtain the stability estimate:

$$\max_{mN\leq k\leq(m+1)N-2}\left\|A^{\frac{1}{2}}\frac{u_{k+2}-2u_{k+1}+u_k}{\tau^2}\right\|_H+\max_{mN+1\leq k\leq(m+1)N}\left\|A\frac{u_k-u_{k-1}}{\tau}\right\|_H \tag{25}$$

$$+\frac{1}{2}\max_{mN+1\leq k\leq(m+1)N}\|A^{\frac{3}{2}}u_k\|_H\leq(2+\tau|b|(N-2))^m b_0$$

$$+\sum_{j=1}^{m}(2+\tau|b|(N-2))^{m-j}\tau\sum_{s=(j-1)N+1}^{jN}\|A^{\frac{1}{2}}f(t_s)\|_H, m=0,1,...$$

for the solution of DS (4). □

3. Applications

Note that the generality of this approach permits studying of a general class of DPDEs. We consider the applications of Theorem 2 for three types of problems. First, the mixed problem for the one-dimensional DPDE with nonlocal conditions:

$$\begin{cases}\frac{\partial^3 u(t,x)}{\partial t^3}-(a(x)u_{tx}(t,x))_x+\delta u_t(t,x)\\ =b\left(-(a(x)u_x(t-w,x))_x+\delta u(t-w,x)\right)+f(t,x),\\ 0<t<\infty,\ 0<x<l,\\ u(t,0)=u(t,l),\ u_x(t,0)=u_x(t,l),\quad 0\leq t<\infty,\\ u(t,x)=g(t,x),-w\leq t\leq 0,0\leq x\leq l\end{cases} \tag{26}$$

is studied. Under compatibility conditions, Problem (26) has a unique solution $u(t,x)$ for the given smooth functions $a(x)\geq a>0,\ x\in(0,l),\ \delta>0,\ a(l)=a(0),\ g(t,x),\ -w\leq t\leq 0,\ 0\leq x\leq l,\ f(t,x),\ 0<t<\infty,\ 0<x<l,$ and $b\in R^1$.

The construction of full discretization to Problem (26) is completed in two stages. In the first stage, we consider the uniform grid space:

$$[0,l]_h=\{x=x_n: x_n=nh,\ 0\leq n\leq M,\ Mh=l\}$$

with step $h > 0$. Let $L_{2h} = L_2([0,l]_h)$ be a Hilbert space of the grid functions $\varphi^h(x) = \{\varphi_n\}_0^M$ defined on $[0,l]_h$, equipped with the norm:

$$\|\varphi^h\|_{L_{2h}} = \left(\sum_{x\in[0,l]_h} |\varphi(x)|^2 h \right)^{1/2}.$$

Let A_h^x be the second order difference operator defined by:

$$A_h^x \varphi^h(x) = \{-(a(x)\varphi_{\bar{x}})_{x,n} + \delta\varphi_n\}_1^{M-1} \tag{27}$$

acting in the space of grid functions $\varphi^h(x) = \{\varphi_n\}_0^M$ satisfying the conditions $\varphi_0 = \varphi_M$, $\varphi_1 - \varphi_0 = \varphi_M - \varphi_{M-1}$. It is well known that A_h^x is a self-adjoint positive definite operator in L_{2h}. Applying A_h^x in (26), we can obtain the initial value problem for an infinite system of third order differential equations:

$$\begin{cases} u_{ttt}^h(t,x) + A_h^x u_t^h(t,x) = bA_h^x u(t-w,x) + f^h(t,x), \\ 0 < t < \infty, x \in [0,l]_h, \\ u^h(t,x) = g^h(t,x), -w \le t \le 0, x \in [0,l]_h. \end{cases} \tag{28}$$

In the second stage, we use DS (4) for (28):

$$\begin{cases} \frac{u_{k+2}^h(x)-3u_{k+1}^h(x)+3u_k^h(x)-u_{k-1}^h(x)}{\tau^3} + A_h^x \frac{u_{k+2}^h(x)-u_{k+1}^h(x)}{\tau} \\ = bA_h^x u_{k-N}^h(x) + f_k^h(x), f_k^h(x) = f^h(t_k,x), k \ge 1,\ x \in [0,l]_h, \\ u_k^h(x) = g^h(t_k,x), -N \le k \le 0, \\ (I_h + \tau^2 A_h^x)\frac{u_1^h(x)-u_0^h(x)}{\tau} = g_t^h(0,x), \\ (I_h + \tau^2 A_h^x)\frac{u_2^h(x)-2u_1^h(x)+u_0^h(x)}{\tau^2} = g_{tt}^h(0,x), x \in [0,l]_h, \\ (I_h + \tau^2 A_h^x)\frac{u_{mN+1}^h(x)-u_{mN}^h(x)}{\tau} = \frac{u_{mN}^h(x)-u_{mN-1}^h(x)}{\tau}, \\ (I_h + \tau^2 A_h^x)\frac{u_{mN+2}^h(x)-2u_{mN+1}^h(x)+u_{mN}^h(x)}{\tau^2} \\ = \frac{u_{mN}^h(x)-2u_{mN-1}^h(x)+u_{mN-2}^h(x)}{\tau^2}, m = 1,2,\ldots \end{cases} \tag{29}$$

Theorem 3. *The solutions of DS (29) obey the stability estimates:*

$$\max_{mN\le k\le(m+1)N-2} \left\| \frac{u_{k+2}^h - 2u_{k+1}^h + u_k^h}{\tau^2} \right\|_{W_{2h}^1} + \max_{mN+1\le k\le(m+1)N} \left\| \frac{u_k^h - u_{k-1}^h}{\tau} \right\|_{W_{2h}^2}$$

$$+\frac{1}{2}\max_{mN+1\le k\le(m+1)N} \|u_k^h\|_{W_{2h}^3} \le C_1 \Big[(2+\tau|b|(N-2))^m b_0^h$$

$$+ \sum_{j=1}^m (2+\tau|b|(N-2))^{m-j}\tau \sum_{s=(j-1)N+1}^{jN} \|A^{\frac{1}{2}} f(t_s)\|_{W_{2h}^1} \Big], m = 0,1,\ldots,$$

$$b_0^h = \max\left\{ \max_{-N\le k\le 0} \|A^{\frac{1}{2}} g_{tt}^h(t_k)\|_{W_{2h}^1}, \max_{-N\le k\le 0} \|g_t^h(t_k)\|_{W_{2h}^2}, \max_{-N\le k\le 0} \|g^h(t_k)\|_{W_{2h}^3} \right\}.$$

hold, where C_1 does not depend on $\tau, h, g^h(t_k)$, and $f_k^h(x)$.

Proof. DS (29) can be written in abstract form:

$$\begin{cases} \frac{u_{k+2}^h - 3u_{k+1}^h + 3u_k^h - u_{k-1}^h}{\tau^3} + A_h \frac{u_{k+2}^h - u_{k+1}^h}{\tau} = bA_h u_{k-N}^h + f_k^h, k \geq 1, \\ u_k^h = g_k^h, -N \leq k \leq 0, \\ (I_h + \tau^2 A_h) \frac{u_1^h - u_0^h}{\tau} = g_t^h(0), (I_h + \tau^2 A_h) \frac{u_2^h - 2u_1^h + u_0^h}{\tau^2} = g_{tt}^h(0), \\ (I_h + \tau^2 A_h) \frac{u_{mN+1}^h - u_{mN}^h}{\tau} = \frac{u_{mN}^h - u_{mN-1}^h}{\tau}, \\ (I_h + \tau^2 A_h) \frac{u_{mN+2}^h - 2u_{mN+1}^h + u_{mN}^h}{\tau^2} = \frac{u_{mN}^h - 2u_{mN-1}^h + u_{mN-2}^h}{\tau^2}, m = 1, 2, ... \end{cases} \quad (30)$$

in a Hilbert space L_{2h} with a self-adjoint positive definite operator $A_h = A_h^x$. Here, $g_k^h = g_k^h(x)$, $f_k^h = f_k^h(x)$, and $u_k^h = u_k^h(x)$ are known and unknown abstract mesh functions defined on $[0, l]_h$. Therefore, the estimate of Theorem 3 follows from the estimate (25). Theorem 3 is proven. □

Second, let Ω be the unit open cube in the m-dimensional Euclidean space. $\mathbb{R}^n (x = (x_1, \cdots, x_m) : 0 < x_k < 1, k = 1, \cdots, n)$ with boundary $S, \overline{\Omega} = \Omega \cup S$. In $[0, \infty) \times \Omega$, the mixed problem for the DPDE with the Dirichlet condition:

$$\begin{cases} u_{ttt}(t,x) - \sum_{r=1}^{n} (a_r(x) u_{tx_r}(t,x))_{x_r} = -b \sum_{r=1}^{n} (a_r(x) u_{x_r}(t-w,x))_{x_r}, \\ 0 < t < \infty, x \in \Omega, \\ u(t,x) = 0, \ x \in S, \ 0 \leq t < \infty, \\ u(t,x) = g(t,x), -w \leq t \leq 0, x \in \overline{\Omega} \end{cases} \quad (31)$$

is investigated. Under compatibility conditions, Problem (31) has a unique solution $u(t,x)$ for the given smooth functions $a_r(x) \geq a > 0$, $(x \in \Omega)$, $g(t,x)$, $-w \leq t \leq 0$, $x \in \overline{\Omega}$, $f(t,x)$, $0 < t < \infty$, $x \in \Omega$, and $b \in R^1$.

The construction of full discretization to Problem (31) is completed in two stages. In the first stage, we consider the uniform grid space:

$$\overline{\Omega}_h = \{x = x_r = (h_1 j_1, \cdots, h_n j_n), j = (j_1, \cdots, j_n), 0 \leq j_r \leq N_r, \\ N_r h_r = 1, \ r = 1, \cdots, n\}, \ \Omega_h = \overline{\Omega}_h \cap \Omega, S_h = \overline{\Omega}_h \cap S$$

and introduce the Hilbert space $L_{2h} = L_2(\overline{\Omega}_h)$ of the grid functions $\varphi^h(x) = \{\varphi(h_1 j_1, \cdots, h_n j_n)\}$ defined on $\overline{\Omega}_h$ equipped with the norm:

$$\left\| \varphi^h \right\|_{L_{2h}} = \left(\sum_{x \in \Omega_h} \left| \varphi^h(x) \right|^2 h_1 \cdots h_n \right)^{\frac{1}{2}}.$$

We consider the difference operator A_h^x defined by the formula:

$$A_h^x u^h = - \sum_{r=1}^{n} \left(\alpha_r(x) u_{x_r}^h \right)_{x_r, j_r}, \quad (32)$$

acting in the space of grid functions $u^h(x)$, which satisfy the conditions $u^h(x) = 0$ for all $x \in S_h$. It is well known that A_h^x is a self-adjoint positive definite operator in L_{2h}. Applying A_h^x in (31), we can obtain that:

$$\begin{cases} u_{ttt}^h(t,x) + A_h^x u_t^h(t,x) = bA_h^x u(t-w,x) + f^h(t,x), \\ 0 < t < \infty, x \in \Omega_h, \\ u^h(t,x) = g^h(t,x), -w \le t \le 0, x \in \overline{\Omega}_h. \end{cases} \tag{33}$$

In the second stage, we also get the difference scheme as the one-dimensional problem case:

$$\begin{cases} \frac{u_{k+2}^h(x) - 3u_{k+1}^h(x) + 3u_k^h(x) - u_{k-1}^h(x)}{\tau^3} + A_h^x \frac{u_{k+2}^h(x) - u_{k+1}^h(x)}{\tau} \\ = bA_h^x u_{k-N}^h(x) + f_k^h(x), f_k^h(x) = f^h(t_k,x), k \ge 1, \ x \in \Omega_h, \\ u_k^h(x) = g^h(t_k,x), -N \le k \le 0, \\ (I_h + \tau^2 A_h^x) \frac{u_1^h(x) - u_0^h(x)}{\tau} = g_t^h(0,x), \\ (I_h + \tau^2 A_h^x) \frac{u_2^h(x) - 2u_1^h(x) + u_0^h(x)}{\tau^2} = g_{tt}^h(0,x), x \in \overline{\Omega}_h, \\ (I_h + \tau^2 A_h^x) \frac{u_{mN+1}^h(x) - u_{mN}^h(x)}{\tau} = \frac{u_{mN}^h(x) - u_{mN-1}^h(x)}{\tau}, x \in \overline{\Omega}_h \\ (I_h + \tau^2 A_h^x) \frac{u_{mN+2}^h(x) - 2u_{mN+1}^h(x) + u_{mN}^h(x)}{\tau^2} \\ = \frac{u_{mN}^h(x) - 2u_{mN-1}^h(x) + u_{mN-2}^h(x)}{\tau^2}, x \in \overline{\Omega}_h, m = 1,2,... \end{cases} \tag{34}$$

Theorem 4. *The solution of DS (34) obeys the following stability estimates:*

$$\max_{mN \le k \le (m+1)N-2} \left\| \frac{u_{k+2}^h - 2u_{k+1}^h + u_k^h}{\tau^2} \right\|_{W_{2h}^1} + \max_{mN+1 \le k \le (m+1)N} \left\| \frac{u_k^h - u_{k-1}^h}{\tau} \right\|_{W_{2h}^2}$$

$$+\frac{1}{2} \max_{mN+1 \le k \le (m+1)N} \|u_k^h\|_{W_{2h}^3} \le C_2 \left[(2 + \tau|b|(N-2))^m b_0^h \right.$$

$$\left. + \sum_{j=1}^{m} (2 + \tau|b|(N-2))^{m-j} \tau \sum_{s=(j-1)N+1}^{jN} \|A^{\frac{1}{2}} f(t_s)\|_{W_{2h}^1} \right], m = 0,1,...,$$

$$b_0^h = \max \left\{ \max_{-N \le k \le 0} \|A^{\frac{1}{2}} g_{tt}^h(t_k)\|_{W_{2h}^1}, \max_{-N \le k \le 0} \|g_t^h(t_k)\|_{W_{2h}^2}, \max_{-N \le k \le 0} \|g^h(t_k)\|_{W_{2h}^3} \right\}.$$

where C_2 does not depend on $\tau, h, g^h(t_k)$, and $f_k^h(x)$.

Proof. DS (34) can be written in abstract form:

$$\begin{cases} \frac{u_{k+2}^h - 3u_{k+1}^h + 3u_k^h - u_{k-1}^h}{\tau^3} + A_h \frac{u_{k+2}^h - u_{k+1}^h}{\tau} = bA_h u_{k-N}^h + f_k^h, k \ge 1, \\ u_k^h = g_k^h, -N \le k \le 0, \\ (I_h + \tau^2 A_h) \frac{u_1^h - u_0^h}{\tau} = g_t^h(0), (I_h + \tau^2 A_h) \frac{u_2^h - 2u_1^h + u_0^h}{\tau^2} = g_{tt}^h(0), \\ (I_h + \tau^2 A_h) \frac{u_{mN+1}^h - u_{mN}^h}{\tau} = \frac{u_{mN}^h - u_{mN-1}^h}{\tau}, \\ (I_h + \tau^2 A_h) \frac{u_{mN+2}^h - 2u_{mN+1}^h + u_{mN}^h}{\tau^2} = \frac{u_{mN}^h - 2u_{mN-1}^h + u_{mN-2}^h}{\tau^2}, m = 1,2,... \end{cases}$$

in a Hilbert space $L_{2h} = L_2(\overline{\Omega}_h)$ with self-adjoint positive definite operator $A_h = A_h^x$ by Formula (32). Here, $g_k^h = g_k^h(x)$, $f_k^h = f_k^h(x)$, and $u_k^h = u_k^h(x)$ are known and unknown abstract mesh functions defined on $\overline{\Omega}_h$ with the values in L_{2h}. Then, the estimate of Theorem 4 follows from Estimate (25) and the following theorem. □

Theorem 5. *The solution of the difference elliptic problem: [40]*

$$A_h^x u^h(x) = \omega^h(x), \ x \in \Omega_h; u^h(x) = 0, \ x \in S_h$$

obeys the estimate:

$$\sum_{r=1}^{n} \left\| u^h{}_{x_r x_{\bar{r}}} \right\|_{L_{2h}} \leq C_3 ||\omega^h||_{L_{2h}},$$

where C_3 does not depend on h and ω^h.

Third, in $[0,\infty) \times \Omega$, the mixed problem for DPDE with the Neumann boundary condition:

$$\begin{cases} u_{ttt}(t,x) - \sum\limits_{r=1}^{n} (a_r(x) u_{tx_r}(t,x))_{x_r} + \delta u_t(t,x) \\ = b\left(- \sum\limits_{r=1}^{n} (a_r(x) u_{x_r}(t-w,x))_{x_r} + \delta u(t-w,x) \right), \\ 0 < t < \infty, x \in \Omega, \\ \frac{\partial u(t,x)}{\partial p} = 0, \ x \in S, \ 0 \leq t < \infty \\ u(t,x) = g(t,x), -w \leq t \leq 0, x \in \overline{\Omega} \end{cases} \tag{35}$$

is investigated. Here, $\overrightarrow{p}$ is the normal vector to S. Under compatibility conditions, Problem (31) has a unique solution $u(t,x)$ for the given smooth functions $a_r(x) \geq a > 0$, $(x \in \Omega)$, $g(t,x)$, $-w \leq t \leq 0$, $x \in \overline{\Omega}$, $f(t,x)$, $0 < t < \infty$, $x \in \Omega$ and $b \in R^1$.

The construction of full discretization to Problem (35) is completed in two stages. In the first stage, we introduce the second order difference operator A_h^x defined by:

$$A_h^x u^h = - \sum_{r=1}^{n} \left(\alpha_r(x) u_{x_r}^h \right)_{x_r, j_r} + \delta u^h, \tag{36}$$

acting in the space of grid functions $u^h(x)$ that satisfy the conditions $D^h u^h(x) = 0$ for all $x \in S_h$. Here, D^h is the approximation of operator $\frac{\partial}{\partial \overrightarrow{p}}$. It is known that A_h^x is the self-adjoint positive definite operator in L_{2h}. Using the difference operator A_h^x, we get the initial value problem (33). Therefore, in the second stage, we use DS (4) for Problem (33):

$$\begin{cases} \frac{u_{k+2}^h(x) - 3u_{k+1}^h(x) + 3u_k^h(x) - u_{k-1}^h(x)}{\tau^3} + A_h^x \frac{u_{k+2}^h(x) - u_{k+1}^h(x)}{\tau} \\ = bA_h^x u_{k-N}^h(x) + f_k^h(x), f_k^h(x) = f^h(t_k, x), k \geq 1, \ x \in \Omega_h, \\ u_k^h(x) = g^h(t_k, x), -N \leq k \leq 0, \\ (I_h + \tau^2 A_h^x) \frac{u_1^h(x) - u_0^h(x)}{\tau} = g_t^h(0,x), \\ (I_h + \tau^2 A_h^x) \frac{u_2^h(x) - 2u_1^h(x) + u_0^h(x)}{\tau^2} = g_{tt}^h(0,x), x \in \overline{\Omega}_h, \\ (I_h + \tau^2 A_h^x) \frac{u_{mN+1}^h(x) - u_{mN}^h(x)}{\tau} = \frac{u_{mN}^h(x) - u_{mN-1}^h(x)}{\tau}, x \in \overline{\Omega}_h \\ (I_h + \tau^2 A_h^x) \frac{u_{mN+2}^h(x) - 2u_{mN+1}^h(x) + u_{mN}^h(x)}{\tau^2} \\ = \frac{u_{mN}^h(x) - 2u_{mN-1}^h(x) + u_{mN-2}^h(x)}{\tau^2}, x \in \overline{\Omega}_h, m = 1, 2, ... \end{cases} \tag{37}$$

Theorem 6. *The solution of the difference scheme (37) obeys the stability estimates in Theorem 4.*

Proof. DS (37) can be written in abstract form:

$$\begin{cases} \frac{u_{k+2}^h - 3u_{k+1}^h + 3u_k^h - u_{k-1}^h}{\tau^3} + A_h \frac{u_{k+2}^h - u_{k+1}^h}{\tau} = bA_h u_{k-N}^h + f_k^h, k \geq 1, \\ u_k^h = g_k^h, -N \leq k \leq 0, \\ (I_h + \tau^2 A_h) \frac{u_1^h - u_0^h}{\tau} = g_t^h(0), (I_h + \tau^2 A_h) \frac{u_2^h - 2u_1^h + u_0^h}{\tau^2} = g_{tt}^h(0), \\ (I_h + \tau^2 A_h) \frac{u_{mN+1}^h - u_{mN}^h}{\tau} = \frac{u_{mN}^h - u_{mN-1}^h}{\tau}, \\ (I_h + \tau^2 A_h) \frac{u_{mN+2}^h - 2u_{mN+1}^h + u_{mN}^h}{\tau^2} = \frac{u_{mN}^h - 2u_{mN-1}^h + u_{mN-2}^h}{\tau^2}, m = 1, 2, ... \end{cases}$$

in a Hilbert space $L_{2h} = L_2(\overline{\Omega}_h)$ with self-adjoint positive definite operator $A_h = A_h^x$ by Formula (36). Here, $g_k^h = g_k^h(x)$, $f_k^h = f_k^h(x)$, and $u_k^h = u_k^h(x)$ are known and unknown abstract mesh functions defined on $\overline{\Omega}_h$ with the values in L_{2h} . Therefore, the estimate of Theorem 6 follows from the estimate (25) and the following theorem. □

Theorem 7. *The solution of the elliptic difference problem: [40]*

$$A_h^x u^h(x) = \omega^h(x),\ x \in \Omega_h; D^h u^h(x) = 0,\ x \in S_h$$

satisfies the estimate:

$$\sum_{r=1}^{n} \left\| {u^h}_{x_r \bar{x_r}} \right\|_{L_{2h}} \le C_4 ||\omega^h||_{L_{2h}},$$

where C_4 is independent of h and ω^h.

4. Numerical Results

It is well known that when the analytical methods fail to work properly, the numerical methods for getting partial differential equations' approximate solutions play a vital role in applied mathematics. In the operator approach, constants in theorems can be large; therefore, in this case a nice stability result must be supported numerically. For this reason, it is important to see that for such a type of theoretical result, we need numerical applications when one cannot know concrete values of constants in stability estimates. Therefore, the first order of accuracy DSs for the solution of one- and two-dimensional DPDEs are presented. To solve this problem, a procedure of modified Gauss elimination is applied. The result of the numerical experiment supports the theoretical statements for the solution of these DSs.

4.1. One-Dimensional Problem

First, we consider the mixed problem, with the exact solution $u\,(t,x) = e^{-t}\cos x$,

$$\begin{cases} u_{ttt}(t,x) - u_{txx}(t,x) = -0.1u_{xx}(t-1,x) \\ -2e^{-t}\cos x - 0.1e^{-(t-1)}\cos x, t > 0,\ 0 < x < \pi, \\ u_x(t,0) = u_x(t,\pi) = 0,\ 0 \le t < \infty, \\ u(t,x) = e^{-t}\cos x,\ -1 \le t \le 0,\ 0 \le x \le \pi \end{cases} \tag{38}$$

for the one-dimensional DPDE.

Applying DS (4), we get the following DS:

$$\begin{cases} \dfrac{u_n^{k+2} - 3u_n^{k+1} + 3u_n^k - u_n^{k-1}}{\tau^3} \\ -\dfrac{u_{n+1}^{k+2} - u_{n+1}^{k+1} - 2\left(u_n^{k+2} - u_n^{k+1}\right) + u_{n-1}^{k+2} - u_{n-1}^{k+1}}{\tau h^2} \\ = -0.1\dfrac{u_{n+1}^{k-N} - 2u_n^{k-N} - u_{n-1}^{k-N}}{h^2} - 2e^{-t_k}\cos x_n - 0.1e^{-(t_{k-N})}\cos x_n, \\ t_k = k\tau,\ lN+1 \le k \le (l+1)N-2,\ l = 0,1,...,\ \ 1 \le k \le N-1\,, \\ N\tau = 1,\ x_n = nh,\ 1 \le n \le M-1,\ Mh = \pi, \\ u_n^k = e^{-t_k}\cos x_n,\ \ -N \le k \le 0,\ 0 \le n \le M, \\ \dfrac{u_n^{k+1} - u_n^k}{\tau} = -e^{-t_k}\cos x_n, -N \le k \le 0,\ 0 \le n \le M, \\ \dfrac{u_n^{k+2} - 2u_n^{k+1} + u_n^k}{\tau^2} = e^{-t_k}\cos x_n,\ -N \le k \le 0,\ 0 \le n \le M, \\ u_1^k - u_0^k = u_M^k - u_{M-1}^k = 0, 0 \le k \le N. \end{cases} \tag{39}$$

It can be written as the second order difference problem with matrix coefficients:

$$A\,u_{n+1} + Bu_n + Cu_{n-1} = D\varphi_n,\ 1 \le n \le M-1; u_0 = u_1,\ \ u_M = u_{M-1}. \tag{40}$$

Here and in the future, we put:

$$u_s = \begin{bmatrix} u_s^{lN} \\ \vdots \\ u_s^{(l+1)N} \end{bmatrix}_{(N+1)\times 1},\quad s = n, n\pm 1.$$

Here,

$$A = C = \begin{bmatrix} 0 & 0 & 0 & 0 & \cdot & 0 & 0 & 0 \\ 0 & 0 & a & -a & \cdot & 0 & 0 & 0 \\ 0 & 0 & 0 & a & \cdot & 0 & 0 & 0 \\ \cdot & \cdot & \cdot & \cdot & \cdot & \cdot & \cdot & \cdot \\ 0 & 0 & 0 & 0 & \cdot & 0 & a & -a \\ 0 & 0 & 0 & 0 & \cdot & 0 & 0 & 0 \\ 0 & 0 & 0 & 0 & \cdot & 0 & 0 & 0 \end{bmatrix}_{(N+1)\times(N+1)},$$

$$B = \begin{bmatrix} 1 & 0 & 0 & 0 & 0 & \cdot & 0 & 0 & 0 & 0 \\ b & -3b & 3b-c & b-c & 0 & \cdot & 0 & 0 & 0 & 0 \\ 0 & b & -3b & 3b-c & c-b & \cdot & 0 & 0 & 0 & 0 \\ 0 & 0 & b & -3b & 3b-c & \cdot & 0 & 0 & 0 & 0 \\ \cdot & \cdot & \cdot & \cdot & \cdot & \cdot & \cdot & \cdot & \cdot & \cdot \\ 0 & 0 & 0 & 0 & 0 & \cdot & 3b-c & c-b & 0 & 0 \\ 0 & 0 & 0 & 0 & 0 & \cdot & -3b & 3b-c & c-b & 0 \\ 0 & 0 & 0 & 0 & 0 & \cdot & b & -3b & 3b-c & c-b \\ -\frac{1}{\tau} & \frac{1}{\tau} & 0 & 0 & 0 & \cdot & 0 & 0 & 0 & 0 \\ \frac{1}{\tau^2} & -\frac{2}{\tau^2} & \frac{1}{\tau^2} & 0 & 0 & \cdot & 0 & 0 & 0 & 0 \end{bmatrix}_{(N+1)\times(N+1)},$$

where:

$$a = \frac{1}{\tau h^2},\ b = -\frac{1}{\tau^3},\ c = \frac{2}{\tau h^2},$$

$$\varphi_n = \begin{bmatrix} \varphi_n^{lN} \\ \vdots \\ \varphi_n^{(l+1)N} \end{bmatrix}_{(N+1)\times 1},\ \left\{ \begin{array}{l} \varphi_n^{lN} = \cos x_n,\ -M \le k \le 0,\ 0 \le n \le M, \\ \\ \varphi_n^k = f(t_k, x_n) = -0.1\dfrac{u_{n+1}^{k-N} - 2u_n^{k-N} - u_{n-1}^{k-N}}{h^2} \\ -2e^{-t_k}\cos x_n - 0.1e^{-(t_{k-N})}\cos x_n, \\ \\ t_k = k\tau, lN+1 \le k \le (l+1)N-2, \\ l = 0,1,\cdots,\ 1 \le n \le M-1, \\ \\ \varphi_n^{(l+1)N-1} = -\cos x_n\ -M \le k \le 0,\ 0 \le n \le M, \\ \\ \varphi_n^{(l+1)N} = \cos x_n,\ -M \le k \le 0,\ 0 \le n \le M, \end{array} \right\}$$

and $D = I_{N+1}$ is the identity matrix.

To solve this second order difference problem, we use the following formula:

$$u_n = \alpha_{n+1}u_{n+1} + \beta_{n+1},\quad n = M-1,...,1,0, \tag{41}$$

where $u_M = (I - \alpha_M)^{-1} \beta_M, \alpha_j \ \ (j = 1, ..., M-1)$ are $(N+1) \times (N+1)$ square matrices, $\beta_j \ \ (j = 1, ..., M-1)$ are $(N+1) \times 1$ column matrices, α_1 is the identity, and β_1 is zero matrices, and:

$$\begin{aligned} \alpha_{n+1} &= -\left(B + C\alpha_n\right)^{-1} A, \\ \beta_{n+1} &= \left(B + C\alpha_n\right)^{-1} \left(D\varphi_n - C\beta_n\right), n = 1, ..., M-1. \end{aligned} \tag{42}$$

The errors are computed by:

$$E_M^N = \max_{lN+1 \le k \le (l+1)N-1, 1 \le n \le M-1} \left| u(t_k, x_n) - u_n^k \right| \tag{43}$$

of the numerical solutions, where $u(t_k, x_n)$ represents the exact solution and u_n^k represents the numerical solution at (t_k, x_n), and the results are given in Table 1.

Table 1. Errors of difference scheme (DS) (39).

l/*N*,*M*	20,20	40,40	80,80
$0, t \in [0,1]$	0.0141	0.0068	0.0040
$1, t \in [1,2]$	0.0559	0.0322	0.0172
$2, t \in [2,3]$	0.1346	0.0746	0.0392
$3, t \in [3,4]$	0.2011	0.1011	0.0561

4.2. Two-Dimensional Problem

Second, the mixed problem with the Dirichlet condition:

$$\begin{cases} u_{ttt}(t,x,y) - u_{txx}(t,x,y) - u_{tyy}(t,x,y) = -0.1u_{xx}(t-1,x) - 0.1u_{yy}(t-1,x,y) \\ -3e^{-t} \sin x \sin y - 0.2e^{-(t-1)} \sin x \sin y, \ t > 0, \ \ 0 < x, y < \pi, \\ u(t,0,y) = u(t,\pi,y) = 0, \ 0 \le t < \infty, \ 0 \le y \le \pi, \\ u(t,x,0) = u(t,x,\pi) = 0, \ 0 \le t < \infty, \ 0 \le x \le \pi \\ u(t,x,y) = e^{-t} \sin x \sin y, \ -1 \le t \le 0, \ \ 0 \le x, y \le \pi \end{cases} \tag{44}$$

for the two-dimensional DPDE is considered. The exact solution of Problem (44) is $u\left(t,x,y\right) = e^{-t} \sin x \sin y$.

Applying DS (4) to the problem (44), we get the following DS of the first order of accuracy in t:

$$\begin{cases} \dfrac{u_{n,m}^{k+2}-3u_{n,m}^{k+1}+3u_{n,m}^{k}-u_{n,m}^{k-1}}{\tau^3} \\ -\dfrac{u_{n+1,m}^{k+2}-u_{n+1,m}^{k+1}-2\left(u_{n,m}^{k+2}-u_{n,m}^{k+1}\right)+u_{n-1,m}^{k+2}-u_{n-1,m}^{k+1}}{\tau h^2} \\ -\dfrac{u_{n,m+1}^{k+2}-u_{n,m+1}^{k+1}-2\left(u_{n,m}^{k+2}-u_{n,m}^{k+1}\right)+u_{n,m-1}^{k+2}-u_{n,m-1}^{k+1}}{\tau h^2} \\ =-0.1\dfrac{u_{n+1,m}^{k-N}-2u_{n,m}^{k-N}-u_{n-1,m}^{k-N}}{h^2}-\dfrac{u_{n,m+1}^{k-N}-2u_{n,m}^{k-N}-u_{n,m-1}^{k-N}}{h^2} \\ -3e^{-t_k}\sin x_n \sin y_m - 0.2e^{-t_{k-N}}\sin x_n \sin y_m, \\ t_k=k\tau,\ lN+1\le k\le (l+1)N-2, \\ l=0,1,...,\ \ 1\le k\le N-1\,, \\ N\tau=1,\ x_n=nh,\ y_m=mh\ 1\le n,m\le M-1,\ Mh=\pi, \\ u_{n,m}^{k}=\sin x_n \sin y_m,\ -N\le k\le 0,\ \ 0\le n,m\le M, \\ \dfrac{u_{n,m}^{k+1}-u_{n,m}^{k}}{\tau}=\sin x_n \sin y_m, \\ -N\le k\le 0,\ 0\le n,m\le M, \\ \dfrac{u_{n,m}^{k+2}-2u_{n,m}^{k+1}+u_{n,m}^{k}}{\tau^2}=\sin x_n \sin y_m, \\ -N\le k\le 0,\ 0\le n,m\le M, \\ u_{0,m}^{k}=u_{M,m}^{k}=0,\ 0\le k\le N, 0\le m\le M, \\ u_{n,0}^{k}=u_{n,M}^{k}=0,\ 0\le k\le N, 0\le n\le M. \end{cases} \tag{45}$$

It can be written as the second order difference problem with the matrix coefficients' form:

$$A\,u_{n+1}+Bu_n+Cu_{n-1,}=D\varphi_n,\quad 1\le n\le M-1; u_0=0,\ \ u_M=0. \tag{46}$$

Here and in the future, we put:

$$u_s=\left[u_{0,s}^{lN},\cdots,u_{0,s}^{(l+1)N},u_{1,s}^{lN},\cdots,u_{1,s}^{(l+1)N},\cdots,u_{M,s}^{lN},\cdots,u_{M,s}^{(l+1)N}\right]_{(N+1)(M+1)\times 1}^{T},\quad s=n,n\pm 1.$$

A, B, C, I are $(N+1)(M+1)\times(N+1)(M+1)$ square matrices, and I, R are identity matrices.

Here,

$$a=\frac{1}{\tau h^2},\ b=-\frac{1}{\tau^3},\ c=\frac{2}{\tau h^2},$$

$$A=C=\begin{bmatrix} 0 & 0 & . & 0 & 0 \\ 0 & E & . & 0 & 0 \\ . & . & . & . & . \\ 0 & 0 & . & E & 0 \\ 0 & 0 & . & 0 & 0 \end{bmatrix},$$

$$B=\begin{bmatrix} Q & O & O & O & . & O & O & O & O \\ E & D & E & O & . & O & O & O & O \\ O & E & D & E & . & O & O & O & O \\ . & . & . & . & . & . & . & . & . \\ O & O & O & O & . & 0 & E & D & E \\ O & O & O & O & O & O & O & O & Q \end{bmatrix},$$

where:

$$E=\begin{bmatrix} 0 & 0 & 0 & 0 & \cdots & 0 & 0 & 0 \\ 0 & 0 & a & -a & \cdots & 0 & 0 & 0 \\ 0 & 0 & 0 & a & \cdots & 0 & 0 & 0 \\ \vdots & \vdots & \vdots & \vdots & \ddots & \vdots & \vdots & \vdots \\ 0 & 0 & 0 & 0 & \cdots & 0 & a & -a \\ 0 & 0 & 0 & 0 & \cdots & 0 & 0 & 0 \\ 0 & 0 & 0 & 0 & \cdots & 0 & 0 & 0 \end{bmatrix},$$

$$D=\begin{bmatrix} 1 & 0 & 0 & 0 & \cdots & 0 & 0 & 0 \\ b & -3b & 3b-c & b-c & \cdots & 0 & 0 & 0 \\ 0 & b & -3b & 3b-c & \cdots & 0 & 0 & 0 \\ 0 & 0 & b & -3b & \cdots & 0 & 0 & 0 \\ \vdots & \vdots & \vdots & \vdots & \ddots & \vdots & \vdots & \vdots \\ 0 & 0 & 0 & 0 & \cdots & c-b & 0 & 0 \\ 0 & 0 & 0 & 0 & \cdots & 3b-c & c-b & 0 \\ 0 & 0 & 0 & 0 & \cdots & -3b & 3b-c & c-b \\ -\frac{1}{\tau} & \frac{1}{\tau} & 0 & 0 & \cdots & 0 & 0 & 0 \\ \frac{1}{\tau^2} & -\frac{2}{\tau^2} & \frac{1}{\tau^2} & 0 & \cdots & 0 & 0 & 0 \end{bmatrix},$$

$Q = I_{(N+1)\times(N+1)}$, $O = O_{(N+1)\times(N+1)}$,

$$\varphi_n = \begin{bmatrix} \varphi_{0,n}^{lN} \\ \vdots \\ \varphi_{0,n}^{(l+1)N} \\ \varphi_{1,n}^{lN} \\ \vdots \\ \varphi_{1,n}^{(l+1)N} \\ \vdots \\ \varphi_{M,n}^{lN} \\ \vdots \\ \varphi_{M,n}^{(l+1)N} \end{bmatrix}_{(M+1)(N+1)\times 1}, \left\{ \begin{array}{l} \varphi_{m,n}^{lN} = \sin x_n \sin y_m, \\ -M \le k \le 0,\ 0 \le n,m \le M, \\[2mm] \varphi_{m,n}^{k} = -0.1\dfrac{u_{n+1,m}^{k-N} - 2u_{n,m}^{k-N} - u_{n-1,m}^{k-N}}{h^2} \\[2mm] -\dfrac{u_{n,m+1}^{k-N} - 2u_{n,m}^{k-N} - u_{n,m-1}^{k-N}}{h^2} \\ -3e^{-t_k}\sin x_n \sin y_m \\ -0.2e^{-t_{k-N}}\sin x_n \sin y_m, \\[2mm] t_k = k\tau, lN+1 \le k \le (l+1)N-2, \\[2mm] l = 0,1,\cdots,\ 1 \le n,m \le M-1, \\[2mm] \varphi_{m,n}^{(l+1)N-1} = -\sin x_n \sin y_m, \\ -M \le k \le 0,\ 0 \le n,m \le M, \\[2mm] \varphi_{m,n}^{(l+1)N} = \sin x_n \sin y_m, \\ -M \le k \le 0,\ 0 \le n,m \le M, \end{array} \right\}$$

To solve this second order difference problem, we use the following formula:

$$u_n = \alpha_{n+1}u_{n+1} + \beta_{n+1},\quad n = M-1,...,1,\ u_M = 0, \tag{47}$$

where $\alpha_j\ \ (j = 1,...,M-1)$ are $(N+1)(M+1)\times(N+1)(M+1)$ square matrices, $\beta_j\ \ (j = 1,...,M-1)$ are $(N+1)(M+1)\times 1$ column matrices, and α_1 and β_1 are zero matrices and:

$$\alpha_{n+1} = -\left(B + C\alpha_n\right)^{-1} A_n,$$
$$\beta_{n+1} = \left(B + C\alpha_n\right)^{-1} \left(D\varphi_n - C\beta_n\right), n = 1, ..., M-1. \tag{48}$$

The errors are computed by:

$$lE_M^N = \max_{lN+1 \leq k \leq (l+1)N-1, 1 \leq n,m \leq M-1} \left| u(t_k, x_n, y_m) - u_{n,m}^k \right| \tag{49}$$

of the numerical solutions, where $u(t_k, x_n, y_m)$ represents the exact solution and $u_{n,m}^k$ represents the numerical solution at (t_k, x_n, y_m), and the results are given in Table 2.

Table 2. Errors of the difference scheme (45).

l/N,M	**10,10**	**20,20**	**40,40**
$0, t \in [0,1]$	0.0370	0.0162	0.0083
$1, t \in [1,2]$	0.0840	0.0456	0.0236
$2, t \in [2,3]$	0.1028	0.0543	0.0276
$3, t \in [3,4]$	0.1008	0.0521	0.0261

As seen in Tables 1 and 2, we obtained some numerical results. If M and N are doubled, the values of the errors decrease by a factor of approximately 1/2 for DS (39) and (45), respectively.

5. Conclusions

1. In this paper, the absolutely stable DS of a first order of accuracy for the approximate solution of the DPDE in a Hilbert space was presented. The theorem on the stability of this difference scheme was proven. In practice, stability estimates for the solutions of three-step difference schemes for different types of delay partial differential equations were obtained. Numerical results were given.
2. The mixed problem for the one-dimensional DPDE with the Dirichlet condition was studied in [41]. The first and second order of accuracy DSs for the numerical solution of this problem were presented. The illustrative numerical results were provided. We are interested in studying absolutely stable DSs of a high order of accuracy of the approximate solution of the initial value problem (1) for the DPDE in a Hilbert space.
3. Applying this approach and the method [30], we could study the existence and uniqueness of a bounded solution of the initial value problem for the semilinear DPDE:

$$\begin{cases} u_{ttt}(t) + Au_t(t) = f\left(t, u(t-w)\right), \ 0 < t < \infty, \\ u(t) = g(t), -w \leq t \leq 0 \end{cases} \tag{50}$$

in a Hilbert space H with an unbounded operator A. Moreover, applying the method of [28], we can investigate the convergence of DSs for the numerical solution of Problem (50).

Author Contributions: Investigation, A.A., E.H. and S.I.. All authors read and approved the final version of this manuscript.

Acknowledgments: The authors thank the reviewers and A. L.Skubachevskii (RUDN) and Ch. Ashyralyyev (CU) for their helpful suggestions.

Funding: This research was funded by "Russian Foundation for Basic Research (RFBR) grant number 16–01–00450.

Conflicts of Interest: The authors declare no conflict of interest.

References

1. Gabov S.A.; Sveshnikov, A.G. *Problems of the Dynamics of Stratified Fluids*; Nauka: Moscow, Russian, 1986. (In Russian)
2. Kozhanov, A.I. Mixed boundary value problem for some classes of third order differential equations. *Mat. Sb.* **1982**, *118*, 504–522. (In Russian)
3. Nagumo, J.; Arimoto, S.; Yoshizawa, S. An active pulse transmission line simulating nerve axon. *Proc. JRE* **1962**, *50*, 2061–2070.
4. Amirov, S.; Kozhanov, A.I. Mixed boundary value problem for a class of strongly nonlinear sobolev-type equations of higher order. *Dokl. Math.* **2013**, *88*, 446–448.
5. Apakov, Y. On the solution of a boundary-value problem for a third-order equation with multiple characteristics. *Ukrainian Math. J.* **2012**, *64*, 1–12.
6. Apakov, Y.; Irgashev, B. Boundary-value problem for a generate high-odd order equation. *Ukrainian Math. J.* **2015**, *66*, 1475–1490.
7. Apakov, Y.; Rutkauskas, S. On a boundary value problem to third order pde with multiple characteristics. *Nonlinear Anal. Model. Control* **2011**, *16*, 255–269.
8. Arjmand, D. Highly Accurate Difference Schemes for the Numerical Solution of Third-Order Ordinary and Partial Differential Equations. Master's Thesis, Numerical Analysis at the Scientific Computing, Royal Institute of Technology, Stockholm, Sweden, 2010.
9. Kudu, M.; Amirali, I. Method of lines for third order partial differential equations. *J. Appl. Math.* **2014**, *2*, 33–36.
10. Latrous, C.; Memou, A. A three-point boundary value problem with an integral condition for a third-order partial differential equation. *Abstr. Appl. Anal.* **2005**, *2005*, 33–43.
11. Niu, J.; Li, P. Numerical algorithm for the third-order partial differential equation with three-point boundary value problem. *Abstr. Appl. Anal.* **2014**, *2014*, 630671.
12. Belakroum, K.; Ashyralyev, A.; Guezane-Lakoud, A. A note on the nonlocal boundary value problem for a third order partial differential equation. *Filomat* **2018**, *32*, 801–808.
13. Ardito, A.; Ricciardi, P. Existence and regularity for linear delay partial differential equations. *Nonlinear Anal.* **1980**, *4*, 411–414.
14. Arino, A. *Delay Differential Equations and Applications*; Springer: Berlin, Germany, 2006.
15. Blasio, G.D. Delay differential equations with unbounded operators acting on delay terms. *Nonlinear Anal.* **2003**, *53*, 1–18.
16. Skubachevskii, A.L. On the problem of attainment of equilibrium for control-system with delay. *Dokl. Akad. Nauk* **1994**, *335*, 157–160.
17. Kurulay, G.; Ozbay, H. Design of first order controllers for a flexible robot arm with time delay. *Appl. Comput.* **2017**, *16*, 48–58.
18. Afuwape, A.U.; Omeike, M.O. Stability and boundedness of solutions of a kind of third-order delay differential equations. *Comput. Appl. Math.* **2010**, *29*, 329–342.
19. Baculíková, B.; Dzurina, J.; Rogovchenko, Y.V. Oscillation of third order trinomial delay differential equations. *Appl. Math. Comput.* **2012**, *218*, 7023–7033.
20. Bereketoglu, H.U.; Karakoç, F.A. Some results on boundedness and stability of a third order differential equation with delay. *An. Stiint. Univ. Al. I. Cuza Iasi. Mat. (NS)* **2005**, *51*, 245–258.
21. Cahlon, B.; Schmidt, D. Stability criteria for certain third-order delay differential equations. *J. Comput. Appl. Math.* **2006**, *188*, 319–335.
22. Domoshnitsky, A.; Shemesh, S.; Sitkin, A.; Yakovi, E.; Yavich, R. Stabilization of third-order differential equation by delay distributed feedback control. *J. Inequal. Appl.* **2018**, *341*, doi:10.1186/s13660-018-1930-5.
23. Grace, S.R. Oscillation criteria for a third order nonlinear delay differential equations with time delay. *Opuscula Math.* **2015**, *35*, 485–497.
24. Pikina, G.A. Predictive time optimal algorithm for a third-order dynamical system with delay. *J. Phys. Conf. Ser.* **2017**, *891*, 012278.
25. Xiang, H. Oscilation of the third-order nonlinear neutral differential equations with distributed time delay. *Ital. J. Pure Appl. Math.* **2016**, *36*, 769–782.
26. Wu, J. *Theory and Applications of Partial Functional Differential Equations*; Springer: New York, NY, USA, 1996.

27. Agirseven, D. Approximate solutions of delay parabolic equations with the Drichlet condition. *Abstr. Appl. Anal.* **2012**, *2012*, 682752.
28. Ashyralyev, A.; Agirseven, D. On convergence of difference schemes for delay parabolic equations. *Comput. Math. Appl.* **2013**, *66*, 1232–1244.
29. Ashyralyev, A.; Agirseven, D. Well-posedness of delay parabolic difference equations. *Adv. Differ. Equ.* **2014**, *18*, doi:10.1186/1687-1847-2014-18.
30. Ashyralyev, A.; Agirseven, D. Bounded solutions of semilinear time delay hyperbolic differential and difference equations. *Mathematics* **2019**, *7*, 1163.
31. Poorkarimi, H.; Wiener, J.; Shah, S.M. On the exponential growth of solutions to non-linear hyperbolic equations. *Int. J. Math. Sci.* **1989**, *12*, 539–546.
32. Sinestrari, E. On a class of retarded partial differential equations. *Math. Z.* **1984**, *186*, 223–224.
33. Shah, S.M.; Poorkarimi, H.; Wiener, J. Bounded solutions of retarded nonlinear hyperbolic equations. *Bull. Allahabad Math. Soc.* **1986**, *1*, 1–14.
34. Wiener, J. *Generalized Solutions of Functional Differential Equations*; World Scientific: Singapore, 1993.
35. Ashyralyev, A.; Hincal, E.; Ibrahim, S. Stability of the third order partial differential equations with time delay. In Proceedings of the AIP Conference Proceedings, Mersin, Turkey, 06-09 September 2018; Volume 1997, p. 020086.
36. Ashyralyev, A.; Arjmand, D. Taylor's decomposition on four points for solving third-order linear time-varying systems. *J. Franklin Inst.* **2009**, *346*, 651–662.
37. Ashyralyev, A.; Arjmand, D.; Koksal, M. A note on the taylor's decomposition on four points for a third-order differential equation. *Appl. Math. Comput.* **2007**, *188*, 1483–1490.
38. Ashyralyev, A.; Sobolevskii, P.E. *New Difference Schemes for Partial Differential Equations*; Birkhäuser Verlag: Boston, MA, USA; Berlin, Germany, 2004.
39. Fattorini, H.O. *Second Order Linear Differential Equations in Banach Spaces*; Elsevier: Amsterdam, The Netherlands, 1985.
40. Sobolevskii, P.E. *Difference Methods for the Approximate Solution of Differential Equations*; Izdat Voronezh Gosud University: Voronezh, Russia, 1975.
41. Ashyralyev, A.; Hincal, E.; Ibrahim, S. A numerical algorithm for the third order partial differential equation with time delay. In Proceedings of the AIP Conference Proceedings, Maltepe University, Istanbul, Turkey, 4–8 September 2019; Volume 2183, p. UNSP 070014, doi:10.1063/1.5136176.

Sample Availability: Samples of the compounds are available from the authors.

MDPI
St. Alban-Anlage 66
4052 Basel
Switzerland
Tel. +41 61 683 77 34
Fax +41 61 302 89 18
www.mdpi.com

Symmetry Editorial Office
E-mail: symmetry@mdpi.com
www.mdpi.com/journal/symmetry

www.ingramcontent.com/pod-product-compliance
Lightning Source LLC
LaVergne TN
LVHW070701170726
843515LV00004B/454